Make and Test Projects in Engineering Design

Andrew Samuel

With a Foreword by William Lewis

Make and Test Projects in Engineering Design

Creativity, Engagement and Learning

With 242 Figures

 Springer

Andrew Emery Samuel, BMechE, MEngSci, PhD, DEng Hon
Research Professor
Engineering Design Group
Mechanical and Manufacturing Engineering
University of Melbourne
Victoria 3010
Australia

British Library Cataloguing in Publication Data
Samuel, A. E. (Andrew E.), 1934–
    Make and test projects in engineering design: creativity, engagement and learning
    1. Engineering design 2. Testing 3. Engineering models 4. Engineering education
    I. Title
    620′.0044
ISBN 1852339152

Library of Congress Control Number: 2005925182

ISBN-10: 1-85233-915-2
ISBN-13: 978-1-85233-915-9
Printed on acid-free paper

Printed in the United States of America (MVY)

9 8 7 6 5 4 3 2 1

Springer Science+Business Media
springeronline.com

...for my family and especially to Eva for just being...

Some day, when I'm awfully low,
When the world is cold,
I will feel a glow just thinking of you...

Discovery is seeing what everybody else has seen and thinking what nobody else has thought.
Albert Szent-Györgyi

FOREWORD

Writing in a more relaxed age the French essayist, Montaigne, was sceptical of the value of knowledge divorced from personal experience. He wrote in his essay "On Pedantry":

I do not fancy this acquiescence in second-hand hearsay knowledge: for though we may be learned by the help of another's knowledge, we can never be wise but by our own wisdom.

But life is more complex in the twenty-first century than Montaigne could ever have envisaged. In engineering education, educators and students have to reconcile the thirst for an ever-expanding body of scientific knowledge with the need to engender and develop the crucial attributes of professional life. This is the conundrum: professional engineering endeavour is characterised not by the accumulation of knowledge but by responsible action. Action centred around the identification and solution of technical problems for the long-term benefit of the societies of which the engineers are members.

Efforts at reconciling knowledge and wisdom constitute an on-going dynamic, ultimately enriching the lives of those who participate, those who engage thoughtfully in the dialogue between the resources available to society and the fact that "great things are not done by those who sit down and count the cost of every thought and act" (Daniel Gooch in his eulogy of Isambard Kingdom Brunel).

Engineers continually reach out to test the limits of contemporary possibilities. The laws of science are inexorable, but this did not disconcert the young Whittle who recognised very early on in the development of the gas turbine jet engine that "our targets of performance for [the engine's] compressor, combustion chamber assembly and turbine are far beyond anything previously obtained with these components". It is perhaps not surprising that Whittle gathered around him a design and development team of primarily young engineers. Some of his dealings with older design engineers were very negative; too often years of experience in a particular industry had led to inflexible habits of thought; for them engineering design had become a routine, a result which was the very antithesis of Whittle's programme of invention and innovation.

How then to translate these thoughts into a successful programme of endeavour in an undergraduate engineering course? We, the readers of this book, are fortunate that its author, Professor Andrew Samuel, has turned his attention to the resolution of this conundrum. Professor Samuel has unrivalled experience as educator, author, researcher and practitioner, a person who has reflected on and thought deeply about his life experiences in university and industry, who has published widely, and who outside his professional work has designed and built his own home. So the business of translating ideas into hardware is something with which he is very familiar.

Make-and-Test (MaT) projects afford a unique opportunity for engineering students (and academic staff who should be encouraged to take part and lead by example) to put their talents of creativity and knowledge of physics to the test, to combine sustained analytical thinking with the exercise of creativity and imagination. And the results cannot be finessed. Participants – students and academic staff alike – have to accept responsibility for a demonstrable, tangible and public outcome. In a university career stretching over several decades I have taken part in and witnessed many MaT projects, and observed that successful outcomes were always achieved by a combination of imaginative synthetic thinking, sound physical understanding and unremitting attention to detail – all characteristics of good professional practice.

I commend this book to all those who think, as I do, that engineering education should offer memorable and liberating experiences to those participating in a unique social enterprise. Practising designers too can employ this text as a self-learning resource, something to dip into whenever they feel the need for intellectual rejuvenation, the need to clear away the mental cobwebs accumulated through extended bouts of work on those routine matters of bookkeeping and administration always present to a greater or lesser extent in professional life.

William Lewis,
University of Melbourne
April, 2004

PREFACE: THE NEED FOR THIS BOOK

The problem interested me because it illustrated the kind of invention which I like, one which nobody asked me for– You're not sure there's any good reason for it, but suddenly you begin to question "Can it be done?" And when you ask the question, the solution follows. One of the big tricks in inventing is not so much to invent the "how", but the "what". In other words, the creating of the problem is as big an invention as the solving of the problem– sometimes, a much greater invention.

Jacob Rabinow[1]

You see things; and you say "Why?" But I dream things that never were; and I say "Why not?"

George Bernard Shaw, The Serpent, in *Back to Methuselah*, Act 1, "In the Beginning"

This is a book about inventing and testing of ideas. Of course, ideas don't just happen. They need to be engendered by some appropriate problem or problem situation. My main purpose then is to describe how to generate engaging and motivating problem situations within the skill ambit of young engineering students. It is hoped that many new ideas will be generated by this process. Socrates used maieutics[2] for seeking to expose ignorance and at the same time to inspire a search for new knowledge. The basic purpose of the projects described in this book is to act both as an engaging challenge for students of engineering as well as a maieutic for giving practical birth to latent ideas. Without this maieutic influence the projects might have the empty character of children's play exercises.

Testing of ideas in a practical way challenges not only the validity of the ideas but also the manufacturing skills of the participants. Because the scene for all this inventing and testing sits within the engineering design course at Melbourne,[3] most of the project work is planned around the use of simple materials such as balsa wood, newsprint paper and readily available plastics

---

1 De Simone (1968).
2 Maieutic: serving to bring out a person's latent ideas into consciousness (from Greek, *maieutikas*, to act as a midwife).
3 Samuel (1984).

such as polyurethane foam, for example. Sometimes more exotic materials are used, such as fibre glass–epoxy composites, candle wax, rubber balloons, aerosol deodourant, mousetrap springs and even spaghetti. Yet, in spite of the varied and sometimes unfamiliar nature of these materials, all of these inventions are intended to be well within the ingenuity and skill range of first and second year engineering undergraduates. Occasionally professional engineering societies offer creative challenges to anyone interested enough to be challenged. This book is also addressed to participants in such projects, whatever their background or skill level. In essence then, the material tries to encapsulate the experience of engineering design, from the uncertainty of "Can I tackle this problem?" to the "aha!" experience of the joyous discovery of a suitable solution. Although the major part of the book is devoted to "neatly frozen" case examples from the Melbourne design programme, there are many opportunities and exercises throughout the text to explore new make-and-test (MaT) projects. In fact Jacob Rabinow's words above set the scene for the primary purpose of this book, namely to articulate the process of inventing challenging and interesting MaT design exercises for undergraduate engineering students.

Bernard Shaw's quotation above provides the background for exploring the nature of engineering education. Engineering science provides the basic tools for predicting the behaviour of materials and processes we use for building artefacts. Engineering design uses these basic tools to build new artefacts. A problem intrinsic to engineering education is how to combine the progressive serial thinking needed in engineering science, with the lateral "thinking outside the square" style of thinking needed in the world of design synthesis. In this context, the book is also addressed to engineering educators who may be grappling with this problem.

In the heady days of the industrial revolution and through the early parts of the last century, the arts and crafts of engineers were often acquired through a master–apprentice process of learning. James Watt was apprenticed to his father Thomas Watt as an instrument maker. Michael Faraday was a laboratory assistant to the nineteenth century chemist Sir Humphrey Davey, the discoverer of the element chlorine. Thomas Telford, the nineteenth century civil engineer was an apprentice stonemason and eventually the surveyor of public works. Robert Stephenson was apprenticed to his father George as an enginewright. When academic resources were plentiful and there was time to experiment with learning programmes the master-apprentice learning process translated into the small tutorial style of mentoring. In a world of shrinking academic resources and growing need to justify how we spend these resources, it is prudent to examine such issues as "educational effectiveness" and "educational efficiency". The first of these issues, educational effectiveness or the basic question of "Have we learnt anything useful from this experience ?" may well be addressed at the occasionally perceived to be frivolous nature of MaT projects. The fundamental thesis of this book is that, if properly organised, planned and administered, significant learning takes place during MaT projects. In this context administration does not refer to the simple nuts and bolts of how many students to a group or how many groups to a tutorial class, but to the way in which MaT projects areplanned and executed. Moreover, organisation and planning also embrace the way we elicit the thinking progression and self-awareness of students in their reporting process. This is the main Socratic maieutic embedded in MaT projects.

Educational efficiency is the ratio of some quanta of learning to the educational resources spent in the process. MaT projects set the scene for the more traditional style of design projects, and I conjecture that virtually the same quanta of learning are acquired in those traditional design exercises as in the deceptively simple MaT project. The uninformed may perceive MaT projects as frivolous or even trivial. After all, who in our professional life will ever ask us for the design of a balloon-driven vehicle? The essence of an introductory MaT project is its engaging quality. Experienced educators are well aware of the critical differences between teaching and learning. Teaching is planning and delivering educational programmes. Learning may or may not take place in these programmes. As children develop they acquire knowledge through their various senses. Touch and feel present opportunities for exploring the world around us as infants. Open-ended artwork and colourful constructions in paper and other materials lead the way to further discoveries in preschool. Much learning takes place during a child's progress through to formal school. Once there, however, resource limitations preclude the use of experiential exploration of the world around us. Formal educational programmes have a kind of regimented quality about them. MaT projects are a way of re-engaging the informal exploration of the world through challenging engineering problem situations. Once re-introduced to this type of informal childlike explorations of engineering, formal educational programmes inherit the benefits of engagement. This is the beginning of *thinking like engineers*. At the other end of the value scale MaT projects should not be treated as lifelong learning experiences. Taking them beyond these early introductory design experiences is weighed down with the real danger of identifying the serious goals and value systems of design with these early play-exercises. Engagement with engineering design thinking can occur in brief one-off MaT projects. Taking them beyond that may seriously reduce their longer-term effectiveness.

As an anecdotal frame of reference I recount the tale of two university cultures and their widely different approaches to the teaching of undergraduate engineering design. One culture is arbitrarily titled the Classical Established University (CEU) and the other is equally arbitrarily called the New Age Caring University (NACU).

At CEU design is treated as an adjunct to the *meat and potatoes* of the engineering sciences. Design tasks are often *reheated* or slightly modified versions of design projects that have gone before, with staff and students merely going through the motions according to an established educational formula. This is a metaphor for the pejorative classical definitions of lectures as quasi-educational activities, where the lecturer's notes are merely transcribed into the student's notebook without entering into the minds of either party.

At NACU the design task is embedded in a matrix of various right-brain stretching activities such as finger-painting to jazz, evocation of fantasy situations and an almost evangelical approach to creative behaviour. Yet neither university seeks to examine, in a scientific way, the effectiveness or efficiency of their teaching methods in design. They make no attempt at benchmarking their approaches to those of other institutions. In fact, a key similarity of the vast majority of undergraduate engineering design programmes throughout the world is their lack of comparative evaluation with other courses. Many such courses offer the type of MaT projects described in this book. These projects, as do almost all properly resourced design projects, absorb significant staff and student resources. It would seem reasonable to evaluate the amount of learning to be gained from such projects in relation to the quantity of resources they consume. Yet few if any engineering design programmes carry out such comparative studies.[4]

The primary purpose of this book is to help engineering design educators in planning and delivering MaT projects and to help young engineering students in planning the solution to MaT projects. A secondary, though equally important, purpose is to examine the benefits and costs of MaT exercises for all concerned, the educators as well as the students. In addition, questions are raised (not all of them answered) about the process of learning in engineering design, about problem solving in groups and about communicating engineering design ideas.

Chapter 1 is a brief introduction to design problems and design thinking. Chapter 2 is specifically addressed to the design educator. The chapter explores issues concerned with learning effectiveness and efficiency with special focus on the discipline of engineering design. Although there is a considerable volume of literature on engi-

---

4 Weilerstein (1999).

neering education, some of the issues being re-discovered in this field have been well explored in the greater body of general educational research. There are pointers and references to some of these issues in Chapter 2.

Chapter 3 introduces the genesis of MaT projects, together with notions of creative behaviour, student engagement and a brief taxonomy of MaT projects.

Chapter 4 is a compendium of material properties, with special reference to the unconventional materials and devices used in MaT projects. The focus here is on experimental exploration of material behaviour where data from established sources are not readily available. The experimental evaluations make use of readily available simple household equipment. Apart from documenting material properties and energy sources for MaT projects the intention of this chapter is to encourage students to experiment and to collect their own evidence of material behaviour wherever possible.

Chapter 5 presents documented case studies of MaT projects dealing with static structures, representing a selection from those conducted over a period of 25 years at the University of Melbourne.

Chapters 6 and 7 introduce dynamic MaT projects and documented case examples of a sample set from those conducted at Melbourne.

Chapter 8 concludes with a brief introduction to open-ended style MaT projects and opens the horizon to a substantial set of similar projects, together with some sources for generating similar MaT projects.

Because these types of projects are generally untidy student and staff experiences, with some degree of confusion during progress, the documentations are sanitised versions of reality. In general, engineering design experiences recounted in reports read better than they lived. This feature of MaT projects is very similar to the experiences of professional engineering designers during the early ideation and development phases of "real life" engineering projects. Perhaps an unusual feature of some documented MaT projects presented is those where we have asked our students to develop a mathematical model of their MaT exercise and to predict the behaviour of the specific structure or device being constructed.

The book concludes with a substantial reference list and bibliography as well as appendices with conversion tables and some simple models of structural and dynamic behaviour.

Throughout the text there are many exercises to stimulate engagement and attention. Solutions to selected exercises are offered in the appendix. As well, there are many proposals for suitable MaT projects, with some suggested performance criteria, but complete specifications are left for the user to prepare. This aspect of MaT projects is probably the most challenging and should engage both the educator and the student.

Andrew Samuel
Melbourne, April 2004

ACKNOWLEDGEMENTS

Morris read through the letter. Was it a shade too fulsome? No, that was another law of academic life: it is impossible to be excessive in flattery of one's peers.
David Lodge, *Small World* 1984

Quite apart from the many design teams involved in delivering those sometimes ingenious devices described in this book, there are many colleagues who encouraged, inspired, supported and, at times, consoled me during this project. The "project" I refer to is not only the book, but all the many design exercises that led up to its production. The most inspiring feature of design education is the wonderful and often challenging intellect that shines through countless interactions with design teams. The student participants in my "project" are too numerous to mention individually, but I am eternally grateful for the opportunity to have had the privilege of working with all of them.

There are many colleagues to whom I also owe a debt of gratitude, not only for the encouragement and inspiration, but also for the continued support of the MaT project programme at Melbourne. First and foremost I must acknowledge William (Bill) Lewis, who was mostly responsible for my freedom to pursue these unconventional, and sometimes crazy, design projects. It is hard to pin down the origin of some of the MaT projects described in the book. Some came from personal interests, others were suggested by colleagues, but many came from, or were substantively refined at, coffee table discussions in the Mechanical Engineering Staff Common Room and at University House. For their never-ending enthusiasm and ideas I am indebted to Ken Brown, Collin Burvill, Bruce Field, Ted Grange, Tony Perry, Peter McGowan, Bob Steidel, Craig Tischler, John Weir and several others, whose names have sadly left my increasingly feeble memory.

Even in this brief acknowledgement it would be remiss of me to leave out mention of the inspiration I have gained from yet another source. There was a time in academia, now almost lost in the mist of economic rationalism, when academic staff were able to meditate on the purpose and meaning of their existence. A special agency that aided and supported that process was *McAree-Meikle International*, an organisation deeply committed to poking fun and occasionally nail-biting parody at anyone who took themselves too seriously, including me and several of my "driven" colleagues. For their contributions I am most grateful to Ross McAree, Peter Meikle and their close associates, Mustafa Camel, Dr. H. Jorgen, S. Mantra Prahabu and last, but by no means least, that doyen of neatness and fashion cravat, once voted the most beautiful person on the planet, Harvey Waston.

Andrew Samuel
Melbourne
2004

CONTENTS

GLOSSARY OF COMMONLY USED DESIGN TERMS AND SYMBOLS

"I don't know what you mean by 'glory,' Alice said.
Humpty Dumpty smiled contemptuously. "Of course you don't–till I tell you. I meant 'there's a
nice knock-down argument for you!'"
"But 'glory' doesn't mean 'a nice knock-down arument,'" Alice objected.
"When I use a word," Humpty Dumpty said, in a rather scornful tone, "it means just what I
choose it to mean–neither more nor less."
Lewis Carroll, *Through the Looking-Glass* (1872)

Basic Design Terminology

Unlike Carroll's Humpty Dumpty, design professionals cannot simply invent meanings for words. In 2000, Samuel *et al.* remarked:[5]

> *Engineering designers invest substantial effort in intellectual argument. There are national or international engineering design conferences almost every year. Often there are several such conferences in the same year. There are several journals that provide communication between the engineering design community. Yet we use terminology in the engineering design context rather loosely. The result is that "design language" is treated by practitioners and others in the same way as any other spoken language, namely as context based interpretive reasoning. This rather cavalier approach to the fundamental tool of design, namely "codified reasoning based on language" has tended to relegate design to a soft discipline.*

Although certain words may be imposed on design by necessity of dealing with an eclectic range of design problems drawn from a wide variety of engineering sources, some design terms must be seen as basic and almost inviolable to communication between designers and their clients. An excellent start for identifying the proper meaning and usage of some of these basic design words is the *Oxford English Dictionary* (*OED*). This work is probably the greatest literary achievement in the English language, and is in no small way responsible for English being the second most spoken language in the world. It would therefore seem arrogant to ignore the *OED*'s etymology[6] when using the English language as a means of formal communication. Historically our language (English or design) is organic and is undergoing changes as our life experiences change. Some words die through neglect (*e.g.,* the archaic meaning of *shambles* is a *market place* or

---

5 Samuel *et al.* (2000).
6 Etymology is the account of or facts relating to formation of words and development of their meaning (*OED*).

slaughter house – OED) and others are born to facilitate communication in a changing world (*e.g., Internet, software* and *afterburner* are typical examples of twentieth century word inventions). The *OED* is based largely on common usage that existed in written text in 17th and 18th century literature. In particular early 18th century writings had amazingly numerous word inventions that were intended to make the writer appear more erudite. In writing about the word lists of the time Simon Winchester notes:[7]

> *So, fantastic linguistic creations like "abequitate", "bulbulcitate" and "sulleviation" appeared in these books alongside "Archgrammacian" and "contiguate", with lengthy definitions…*

Communicators in engineering design are equally capable of the odd verbal invention. Perhaps (to this author) the most disturbing example is the reckless use of *ontology*, for what is in reality a taxonomy or classification of words and meanings.[8] Bearing these matters in mind, the only caution to intending design communicators is that if there is a useful and readily available plain English word for whatever one would wish to communicate, then avoid inventing a new one. The following words should be regarded as inviolable engineering design terminology(some have well defined etymology in the *OED*):[9]

- *Design Need*: The primary motivation for a design investigation (usually expressed in the form of a *problem statement* – *e.g.*, *"There are too many automobile accidents during holiday periods"* or *"My arthritic aunt is unable to open her milk carton"*);

- *Design goal*: The primary functional objective of a design (usually expressed in terms of outcomes without reference to embodiment – *e.g.*, *"a means for delivering a payload along a water channel"* rather than *"a boat"*);

- *Design objectives*, or *design requirements*: Desired features or characteristics of a specific design (*e.g.*, *"safe"*, *"reliable or robust"*, *"cheap"*);

- *Constraint*: Mandatory design requirement (*e.g.*, *"the device must weigh less than 5 kg"*, or *"It must be inflammable"*);

- *Restriction*: Flexible design requirement (*e.g.*, *"The lecture theatre should accommodate 200 students"*, or *"The cockpit should accommodate 98% human population size"*);

- *Criterion* (or *criteria* – plural): The scale on which the "fitness for purpose" of the design is measured (*e.g.*, for *cheap* the *criterion* is $,

---

7 Winchester (1999).
8 Samuel *et al.*(2000), Ibid.
9 Refer also to the Design Lexicon at www.mame.mu.oz.au/eng_design/language/language.html.

or whatever monetary unit is in use, for *reliable or robust* the criterion is *mean time to failure* or *mean time to repair*, for *comfortable* the
criterion is the *subjective judgement of a group of end users* of the product).

Criterion is probably the most often misunderstood or mis-used engineering design term. Yet, equally probably, it is the single, most precisely
definable, basic design term in the brief list offered here. Proper use of these
few terms in the early definition stages of a design can yield substantial benefits for the design team.

Commonly Used Symbols

A, a	area, cross-section area, acceleration
C_p, C_v	specific heat (at constant pressure or volume = heat energy needed to raise the temperature of a gas by one degree K)
D, d	diameter
E	Modulus of elasticity, (Young's modulus)
F, f	force, friction factor
g	acceleration due to gravity
I_{zz}	second moment of area, mass moment of inertia
k, K	spring stiffness, kinetic energy
l, L	length dimension
m	mass
p	pressure
r, R	radius
S_U	ultimate tensile strength
S_Y	yield strength
t, T	time, material thickness, temperature
v, V	velocity, electric potential
α	thermal coefficient, angle
δ	small change in quantity, deflection of beam
ε	strain (fractional change in length)
∂, ϕ	angle, twist/unit length
μ	Poisson's ratio
ρ	density
π	ratio of circumference to diameter of circle
σ	direct stress
τ	shear stress

1
INTRODUCTION

The situations you have studied in your texts and laboratories are established situations. The situations I have to deal with in everyday life are emergent. Your thorough understanding of established situations is worthless for the things that are really problems to me in everyday life.
Robert Boguslaw, *The New Utopians*, 1965

Dilbert-Melbourne Age newspaper 22/04/03 [Copyright:United Features Syndicate Inc.] Used with permission.

In this cartoon Adam Scott, the artist inventor of Dilbert, captures the essence of engineering synthesis. Everybody designs sometimes. Nobody designs always.

Suppose we were asked which of these three is the odd one, a waterfall, a buttercup, and a steam locomotive. One answer would be, the locomotive because it alone is man-made, another would be, the buttercup, because it alone is alive. But the third possible answer would also be justifiable: the buttercup and the locomotive show evidence of design, but the waterfall does not - its shape simply happens, it has no symmetry, it is not contrived to have any function or to serve any end.[1.1]

The above mental hook was used by Michael French in his introduction to the *Designed World in Invention and Evolution; Design in Nature and Engineering*. Most creative writers are well aware of the value of mental hooks for gaining the attention of their audience before delving into the details of a story. The spirit of this book is to act as a mental hook into the act of designing. One might wonder why we need a mental hook or attention grabber for an undergraduate programme of learning when no such

[1.1] French (1988).

1

attention grabber is necessary (or is it?) in (say) thermodynamics or structural mechanics. The answer might be that although engineering science learning is a natural progression from the sciences of high school, engineering design is a totally new experience. Most of us enter a university course in engineering well prepared for analysis, but totally unprepared for synthesis. Young preschool children have well-developed skills in synthesis while playing with building blocks, playdough and even mud. They often live in the world of fantasy and imagination. The formal training of high school has a serious blunting effect on these creative synthetic skills. Fantasy and imagination are difficult skills to test in the formal examinations that prepare students for university entry. In engineering we need to resharpen these skills so essential to the world of design.

Another reason for the mental hook to the act of designing is articulated in Herbert Simon's introduction to the science of design in *Sciences of the Artificial*:[1.2]

> *In view of the key role of design in professional activity, it is ironic that in this century the natural sciences almost drove the sciences of the artificial from professional school curricula, a development that peaked about two or three decades after the Second World War. Engineering schools gradually became schools of physics and mathematics; medical schools became schools of biological sciences; business schools became schools of finite mathematics. The use of adjectives like "applied" concealed, but did not change the fact...It did not mean that design was continued to be taught, as distinguished from analysis.*

Perhaps a practical attention grabber for intending designers might be to consider what designers do and the extent of their influence. This question is akin to asking what numbers do in arithmetic or in number theory. To begin answering the question we need to establish some early building blocks to the whole human process of designing. All artefacts (man-made products such as can openers, locomotives, housing estates and nuclear power plants) are influenced by designers. All of these artefacts have been generated by some human need reinterpreted by the designer into technical goals. In an ideal world we start with the goals and the designer plans the artefacts to meet the required goals. Of course, this is a vast oversimplification. Depending on the scale of the artefact (can opener or nuclear power plant) there may be many stages of planning involved. But designing and designers are involved in all of them. Additionally, designers not only plan artefacts but also plan systems and organizations. Their (our) influence extends over virtually all human activity from baking a chocolate fudge cake to putting a man on the moon.

1.2 Simon (1996).

Doctors heal people, business managers improve the operation of a business and designers plan. Time-scales of outcomes seriously influence the way we respond to these tasks. Clearly, we see doctoring having an immediate response to our ills. At the other extreme of response time, design plans and their hardware realisations may only be experienced long after the designers creative work is done. In a world of instant almost everything, this time delay can seriously undermine undergraduate enthusiasm about design. MaT projects offer an opportunity to experience the full excitement of design planning and realisation in the short term. A meeting of design educators, inventors, industrialists and government agencies was held in late 1966 at Woods Hole, Cape Cod.[1.3] The meeting was sponsored by the National Academy of Engineering, the National Science Foundation and the U.S. Department of Commerce. The aims of this meeting were:

...to examine the creative processes of invention and innovation, the opportunities for encouraging creative activities in engineering schools, and the possibilities for developing and supporting creative engineering education.

It is reasonable to assume that this meeting was a response to the climate of educational concerns expressed by Herbert Simon's words quoted above. Whatever the reason for this meeting, its influence on engineering education in the United States was substantial. Typical examples are the courses in design offered at Stanford university, the University of California–Berkeley, MIT and Carnegie-Mellon University.[1.4] All of these schools have a considerable history of offering MaT projects to undergraduates. The construction of "bridges" from balsa wood, styrofoam and paper are based in the industrial design programme originally offered by Bob McKim and Jim Adams at Stanford, as is the walking-on-water exercise, irreverently referred to as the "Jesus-Boots" project. The origins of the mouse trap-driven vehicle exercise is attributed to MIT.[1.5] In the introduction to *Education for Innovation*, De Simone[1.6] wrote:

...the art of creative engineering has been orphaned in the engineering schools. ...there has developed...a regrettable and unnecessary schism between the realms of science and engineering. Paradoxically, in the schools of engineering, the art of engineering has been largely neglected.

and later on

But most important, a student learning engineering should be permitted to behave like an engineer.

1.3 De Simone (1968).
1.4 See, for example, McKim (1972,1978); Adams (1974); Fuchs and Steidel (1973); Vesper (1975).
1.5 Fuchs (1976).
1.6 *Ibid.*

At the same meeting, William Bollay of Stanford University noted:

I think it is very important that the student has a chance to exercise his [her] ingenuity in the solution of a real problem, to which he [she] does not know the solution.

The expression of these ideas illustrates the nature of concerns among design educators thirty five years ago. It is a feature of the engineering educational system that these problems are as relevant today as they were then. The intention of this book is to formalise the design and development of MaT projects and to encourage all engineering educators to participate in planning their own versions of MaT exercises. An important caution of which all design students should be aware early in their studies is that although designers propose (plans and possibly prototypes), it is often the holders of the purse strings who dispose the funds required for development of ideas into artefacts. In this context the occasion arises when technical decisions are guided by economic rationale. There are some salutary examples of how economic rationalism can seriously negate a perfectly good design plan. This may well be referred to as "fumbling the future" and snatch abject economic failure from the jaws of technical victory.

One example is the case of Xerox,[1.7] a large and successful corporation, whose management sought to invest in basic idea generation and creativity. This was a thoughtful, forward-looking, decision by a visionary management in times of plenty. They hired some young and creative staff (let's call them *young Turks*)and asked them to create and explore new ideas, or technologies, without any constraints on the directions or financing of their creations. Among other things, these young Turks invented a radically new (at the time) personal computer called the *Alto*. It had some features of the future Macintosh, years ahead of its time. Yet, by the time they asked management for funds to further develop the Alto, the corporation was experiencing a downturn in business. This downturn resulted in an economic battening down of all the hatches of the corporate vessel and important funds, needed for developing the Alto, were denied. Management reasoned (rationalised) that the development of the Alto was not part of Xerox's core business. Even more importantly some young Turks, involved in the early development of the Alto, went on to work for Apple during the development of the Macintosh.

Another example is the case of the ill-fated DC10 developed by the McDonnell Douglas corporation in the race for the first successful, wide-bodied, passenger airliner. The three entrants in the race were Boeing (747), Lockheed (TriStar) and McDonnell Douglas (DC10). By the time McDonnell Douglas had entered the race in 1965, they were more than a

1.7 Smith and Alexander (1988).

year behind the other two contenders. Financial pressures and difficulties with recruiting skilled labour caused management to take unprecedented shortcuts in development against the judgement of their designers.[1.8] Two major air disasters in which DC10 aircraft were involved happened in March 1974 near Paris and in May 1979 at Chicago O'Hare airport. Until that time these two were the worst aircraft accidents of all time. About the Paris air-crash the National Transport Safety Board's *Aircraft Accident Report* 8/76 noted:

> *Finally, although there was apparent redundancy of the flight control systems, the fact that the pressure relief vents between the cargo compartment and the passenger cabin were inadequate and that all the flight control cables were routed beneath the floor placed the aircraft in grave danger in the case of any sudden depressurization causing substantial damage to that part of the structure. All these risks had already become evident, nineteen months earlier, at the time of the Windsor accident, but no efficacious corrective action had followed.*

About the Chicago O'Hare airport accident NTSB wrote in *Aircraft Accident Report* NTSB-AAR-79-17:

> *Contributing to the cause of the accident were the vulnerability of the design of the pylon attach points to maintenance damage; the vulnerability of the design of the leading edge slat system to the damage which produced asymmetry; deficiencies in FAA surveillance and reporting systems which failed to detect and prevent the use of improper maintenance procedures.*

In both reports the investigations focused on weaknesses in design. If one reads the history of the DC10 as recounted by Eddy *et al.*,[1.9] an impression is gained that it seemed almost inevitable that design shortcuts were required to meet the critical deadlines set for production by McDonnell Douglas management. Yet all these deadlines were driven by the economics of highly competitive civil aircraft manufacture.

One of the major benefits of MaT projects is that the designers make all their own decisions about construction and development. It is probably the only time in their careers that they will be able to do so. An important cautionary caveat about MaT projects is that they tend to absorb substantial educational resources. In general they act as very useful attention grabbers for early design studies in engineering. Their longer-term educational benefits have not been studied in any formal way, but there is consistent anecdotal reporting of experiences with "hands-on" types of project work.[1.10] In

1.8 Eddy et al. (1976).
1.9 Op. cit.
1.10 Klenk ,et al. (2002); Biden and Rogers (2002); Egelhoff and Odom (1999); Cottrell and Ressler (1997); Ambrose and Amon (1997); Durfee (1994); Love and Dickerson (1994); Niku (1995); Hibbard and Hibbard (1995); Abdel-Rahman and Shawki (1989).

this context the often quoted benefits are "interest and excitement of both staff and students involved." Although the warm feelings expressed about these types of projects all fall under the general heading of engagement, as previously noted, care in planning and management can bring even greater and perhaps longer term benefits to MaT projects. These benefits are remarkably helpful in the introductory parts of the course, however, they do take time away from the more formal (and often equally creative) elements of design education. Bearing this in mind it is instructive to revisit the objectives and evaluations of engineering design education in universities. Chapter 2, specifically aimed at the design educator, discusses some of these issues in detail.

Because MaT projects rely heavily on team problem solving and visual communication, these topics require some special attention. Clearly, the evolution of the MaT project is best engendered in the guise of some problem situation. Many planners of MaT projects use "fairy-tale" scenarios for setting up the problem situation and the problem boundaries. One could conjecture a twofold rationale for this approach. First of all, a fairy-tale problem situation may be readily accepted, especially by one's more seriously minded academic colleagues, as an appropriate precursor to an apparently trivial design exercise. Secondly, the fairy-tale scene lends itself not only to provocatively unconventional engineering materials, such as elastic bands and spaghetti, but also to such magically staged problem boundaries as the confines of the engineering design office, or the limitations of a shallow reflection pool for water bound vehicles. The Melbourne projects described in the case examples in this book have no such fairy-tale scenarios. In general, they were planned and set as simple challenges that might be solved in a relatively short time, because the time-scale of the Melbourne MaT project is typically 12 student contact hours, spaced out over a four-week period.

The process of problem solving is a serious study. It is expected that the process involves all of the senses and builds on the sum total of all our experiences. The general approach to design problem solving at Melbourne is based on the Socratic dialogue.[1.11] The term "maieutic" refers to a method of teaching by question and answer, as used by Socrates to elicit truths from his students. The word "maieutic" pertains to the obstetric idea of midwifery in bringing forth the birth, not of human tissue, but of ideas. A key feature of the *Socratic maieutic*, applied to solving engineering design problems associated with MaT projects, is that of avoiding the free rein of inexperienced intuitions. Because young and inexperienced engineering designers have yet to build up their store of experience, intuitive design for them

---

1.11 Samuel and Lewis (1986).

may be seriously misleading. At the early introduction to the MaT project problem statement, students are carefully advised of the need to withhold early judgement of the various ideas that may present themselves as solutions. Moreover, as the project is invariably tackled in groups, the group members are also invited to act as a team. The difference is drawn here between a group of problem solving individuals, with individual tastes and inclinations, and a team with shared goals and preferences. Groups of football players just want to handle the ball, whereas a team tries to ensure that goals are kicked. Later in the book further and more substantial reference is made to the nature of problem solving in teams, but in general, MaT project teams are left to work out their own approach to human interactions.

One specific advice given to these teams of problem solvers is to keep records of their various ideas on the way to their solution. This activity is critically influenced by visual communication skills. Communication of ideas can take place in many and varied ways. Engineers, in general, tend to favour sketching as a means of communicating ideas of artefacts at the early development stage. Young engineering students have limited experience or skills in sketching and the use of fine-point writing instruments is a limiting feature of communication in the early years of engineering training. In addition, although sketching may be part of some art programmes in schools, prior to entry to university courses in engineering few students have been exposed to one- and two-point perspective. Figure 1.1 illustrates one- and two-point perspective sketches of a cube. Figure 1.2 shows some simple generic shape models that might be used for sketching exercises to enhance sketching skills A great variety of simple and complex shapes may be built from such generic shapes. Figure 1.3 shows a way one can build complex shapes using perspective sketching and combining generic shapes.

Introducing simple sketching to undergraduates enhances interpersonal communication skills during design planning and idea exploration. However, not all problem solvers can take advantage of improving sketch-

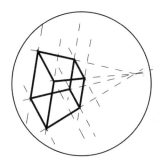
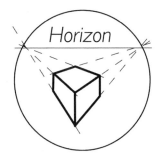

Figure 1.1 One-point and two-point perspective sketches of a cube

Figure 1.2 Simple generic models for sketching exercises

Figure 1.3 Two-point perspective built from generic components

ing skills. Some are more at home with verbal skills and with fluent enumeration of design alternatives at times making use of stick figures. Whatever the approach, the essence of unfettered creative flow of ideas is some comfortable means of communicating these ideas. The following illustrate some of the sketches produced by students involved in MaT projects.

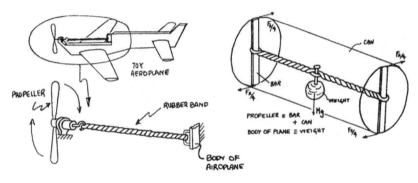

Figure 1.4 Preliminary idea sketches for a rubber-band powered device

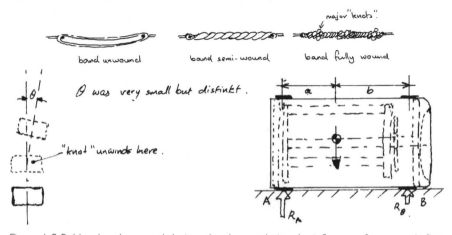

Figure 1.5 Rubber-band powered device; sketches exploring the influence of uneven winding of the rubber drive

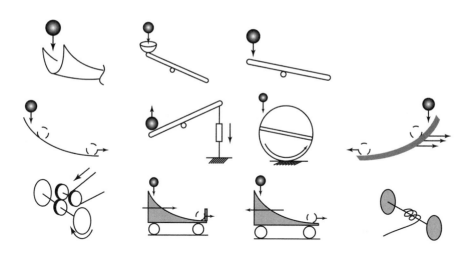

Figure 1.6 Morphological sketches of ideas for a golf ball driven vehicle

Explanation:
Row 1 deals with methods of energy capture from the falling golf ball;groove;
Row 2 deals with ideas of energy conversion;
Row 3 deals with alternative drive embodiment issues;

Figures 1.4 and 1.5 show student designers' idea sketches for a rubber-band powered drive system designed to roll a beverage can along a smooth test track. Figure 1.6 is a morphological[1.12] sketch for a vehicle powered by the potential energy derived from a golf ball at 2 m vertically above the vehicle at the starting line. Figure 1.7 shows further idea sketches presented for the same golf ball driven device by a different student team.

Idea generation is a process closely linked to a person's world view and personal development through experiences. This is particularly the case when we are severely time-constrained in generating ideas, as is the case with short time-scale MaT projects. An amusing example of how personalised idea generation manifests itself is seen in Graham Greene's novel *Our Man in Havana*. In that novel a vacuum cleaner salesman, Jim Wormold, is recruited by British intelligence to report on the suspected rise of a military dictatorship in Cuba. The arrangement is to the advantage of both parties. MI5 needs intelligence and Wormold needs the money generated through his reports. In the absence of real intelligence to report, Wormold resorts to sending sketches of mythical military installations, in the Cuban mountains,

1.12 Morphology is defined by the OED as the study of the form of things. In engineering design this has been interpreted as a study of the form of various attributes of a design. See, for example, Pahl and Beitz (1996); Samuel and Weir (1999).

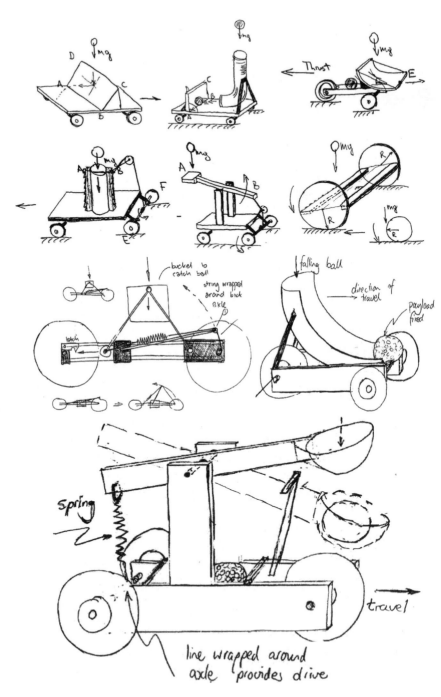

Figure 1.7 Idea sketches for a potential energy-driven device (a golf ball dropped from a height of 2 m vertically above the device)

all having the appearance of very large scale vacuum cleaner parts. Popper also remarks on the nature of belief-systems that grow from unfounded support of ideas:[1.13]

> I found...admirers of Marx, Freud and Adler, were impressed by a number of points common to these theories, and especially by their apparent "explanatory" power. ...The study of any of them seemed to have the effect of an intellectual conversion or revelation, opening our eyes to a new truth hidden from those not yet initiated. Once your eyes were thus opened, you saw confirming instances everywhere: the world was full of "verifications" of the theory. Whatever happened always confirmed it... A Marxist could not open a newspaper without finding on every page confirming evidence of his interpretation of history...

Early development of our view of the world around us and the influences of our environment and background all have substantial impact on our ideas of how things should or do work. I conjecture that many entrants to engineering courses of (say) 30 years ago would have had experiences with automobile engines, bicycle mechanics and Meccano[1.14] style construction kits. Today's engineering entrant is less experienced with such hands-on engineering, but is considerably more informed about computer technology and fantasy computer games. The lack of hands-on experience is limiting, but the suspended judgement needed to operate fantasy computer games is creativity enhancing. Further reference is made to metrics of creative behaviour in Chapter 2, but as a basic introduction to idea generation in MaT projects we need to provide guidance for students. This is especially important for problem solving as part of a team.

As an instructive exercise in group problem solving try to explore a puzzle with a small group of people. This experience will be substantially different in a setting of a group of strangers, when compared to the same puzzle being tackled by a group that know each other well. There is a general reluctance of people to expose their thoughts and thinking processes in front of strangers. Some of the evidence for this type of behaviour is reviewed in Chapter 2. Our approach at Melbourne (this same type of approach is also used at many other universities) has been to expose students to the conceptual blocks, described by Adams,[1.15] and in particular to a game called "barnyard". In this game all students are asked to imagine themselves as barnyard animals (cows go "moo", chickens go "paak", turkeys go "goble" etc.) and to make the appropriate animal sounds while facing their neighbour. The purpose is to expose them all to feeling slightly

1.13 Karl Popper (1963), p 34.
1.14 *Meccano* is a metal construction kit patented by Frank Hornby in 1901, for a brief history see freenet.edmonton.ab.ca/meccano.
1.15 Adams (1974).

ridiculous, thereby (hopefully) freeing up some personal inhibitions. Adams even suggests playing barnyard in much more public places, like street corners or traffic lights, to further enhance the removal of built-up social inhibitions.

Another key feature of problem solving in groups is the generation and judgement of ideas. For free unfettered flow of ideas, no matter how unorthodox they might be, it is essential to delay judgement. de Bono[1.16] recognises this in his approach to group problem solving. In *Six Thinking Hats* two special hats are identified as useful for group problem solving. *Yellow hats* are worn (metaphorically speaking) by the unfettered idea generators. They will not judge any idea, no matter how "way-out" it may seem. This is delayed judgement. *Black hats*, on the other hand, are worn (again as metaphor only, reprising the notion of a judge dealing out capital punishment) by those who make judgement. In this context, a further advice to those new to group problem solving is to continue "wearing the yellow hats" until all ideas are exposed.

A useful metaphor for the free-flowing idea generation aspect of MaT projects is the "false mirror" (*Le Faux Mirroir*) of the artist Rene Magritte. As Samuels and Samuels[1.17] note in their book *Seeing With The Mind's Eye*:

> *This image is a logos for visualisation. The picture shows inner and outer worlds inseparably joined. One has the feeling of looking inward, of looking inside another person; at the same time one has the feeling of looking outward at the world. Suddenly one becomes the person in the picture, staring into and out of his own eye. The superposition of the two views conveys the feeling of seeing with the mind's eye - a feeling of unity, insight and understanding.*

Figure 1.8 Rene Magritte: Le Faux Mirroir, 1928, Museum of Modern Art, New York. Reproduced here with the kind permission of Viscopy Pty. Ltd. on behalf of the artist.

1.16 de Bono (1985).
1.17 Samuels and Samuels (1975).

2
INVENTION, CREATIVITY, ENGAGEMENT: MaT PROJECTS

As an inventor Leonardo was an astonishing genius. Although he lived over 450 years ago he foresaw the coming of advanced technology, and filled his notebooks with thousands of drawings for new machines and weapons, many of which anticipate twentieth century engineering techniques. We see him designing armoured tanks, steam guns, ballistic missiles, flying machines, parachutes, helicopters, underwater diving suits, water turbines, movable cranes, lifting jacks, gearboxes. Each of his inventions is a continuous source of wonder and excitement, displaying both Leonardo's awesome intelligence and an incredible anticipation of the future. Charles Gibbs-Smith and Gareth Reese, The inventions of Leonardo Da Vinci, 1978

Genius is one percent inspiration, ninety-nine percent perspiration.
Thomas Alva Edison, *Harper's Monthly Magazine*, September 1932

Leonardo Da Vinci has become a legendary figure for art and science historians. What makes him so interesting to twentieth century (and earlier) historians? The *Mona Lisa* is another icon of our current association with Leonardo. Although not necessarily regarded as a great painting, it basks in the reflected greatness of Leonardo. More to the point, why did we have to wait till the twentieth century to experience mechanical flight when Leonardo had a sketch invention of a helicopter 450 years ago? The clue to all this is that Leonardo was a great thought experimenter. In more recent history, Ernst Mach, the nineteenth century Austrian physicist and Albert Einstein are the names most often associated with *gedanken*[2.1] experiments, Leonardo certainly foreshadowed them by several centuries. A distinctive feature of Leonardo's work was his extensive use of sketches to develop his ideas. This is not a particularly compact way of exploring scientific principles, but Leonardo did not have the advantage of modern mathematics at his disposal. We are of course fortunate indeed to have inherited Leonardo's many hundreds of beautiful sketches and descriptions of his inventions. They are considerably more accessible than more compact mathematical expressions. The following sketches are part of a collection housed in the Museum of Science and Technology in Milan. A section of this museum is dedicated to working models built from the sketches of Leonardo. Although the sketches range over human and animal anatomy, war machinery, clockwork and architectural plans, the ones chosen as illustrations here are from his work on flying machines with specific reference to wings directly attached to the flyer's body (*ornithopters*).

2.1 *Gedanken* (German): thoughts, ideas; see also Brown (1991); Brown (2002).

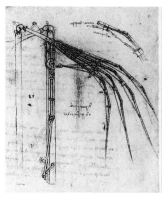

Figure 2.1 Drawing from Codex Atlanticus, *folio 844. The folio, which may be dated to 1496, contains the study for an articulated wing and details of the joints and springs to be used in its construction. The following annotation was made under the large drawing of the wing, "dove-tailed linen cloth", although further down and along the margin Leonardo wrote the remark (translated): "as for the spring mechanism, take thin, hardened wire; if the wire sections between the joints are of the same thickness and length and if each spring has the same number of wire sections the springs thus obtained will be equally strong and resistant."*

Reproduced with the permission of Biblioteca Ambrosiana.

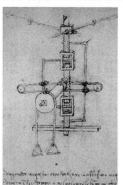

Figure 2.2 Wing drawing from Codex Atlanticus, *folio 858 r. This mechanism was designed to drive the wing via a crank, which wound a rope around a winch.*

Reproduced with the permission of Biblioteca Ambrosiana.

Figure 2.3 Wing drive mechanism from Manuscript B, folio 77 r. The drum is rotated by the thrust provided by the flier's feet placed in the stirrups. This results in the up and down motion used for operating the wing. Leonardo tried to reproduce the mechanism of the joints in a bird's wing during various phases of flight. The model shown is accompanied by a detailed description of the position of the flier's arms and legs, and of the synchronised movements required for the wing to tilt via a system of cords and joints.

Reproduced with the permission of Biblioteca Ambrosiana.

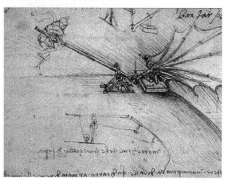

Figure 2.4 Wing lift testing apparatus, from Manuscript B, folio 88 v dating to 1485. Leonardo believed that, if the person testing the device managed to push down the long lever fast enough, the wing connected to it would touch down on the air and lift the wooden block attached to the wing.

Reproduced with the permission of Biblioteca Ambrosiana.

The first sketch is of an articulated wing structure. Leonardo designed this device, complete with traction and torsion mechanisms in the outer part of the wing, to faithfully reproduce the structure of a bird's wing. The purpose of this device was to guarantee the automatic return motion of the flexed wing. Special care was placed on the study of the springs and joints connecting the various sections of the wing.

Perhaps the most illuminating example of Leonardo's approach to his thought experiments about flight is his sketch of a wing testing apparatus shown in Figure 2.4. Having verified that the wing span of a duck is approximately equal to the square root of the weight of the animal, Leonardo estimated the wing span required to lift a man and his flying machine, weighing an overall 400 lbs (136 kg), must be equal to about the square root of the total weight. Hence, Leonardo calculated that the size of the wing should be 12 metres long and 12 metres wide and that two such wings could lift the man and his flying machine into the air and keep them in flight.

In essence, a thought experiment is the reasoning process which responds to questions of the "what-if" type. Einstein's *gedanken* were associated with questions of relativistic time behaviour. For example, one could ask questions such as:

What would a clock appear to show to an observer travelling, at the speed of light, away from (or towards) the clock? What would the observer see when travelling at a speed greater than the speed of light away from (or towards) the clock?

These thought experiments are the easy-to-digest metaphors for the evaluation of questions relating to the behaviour of time at or above the speed of light. The responses elicit great insight into the behaviour of time in a relativistic world. Certainly one could not carry out such experiments physically with current technology.

Similarly, Leonardo could do no more than use thought experiments, albeit with clear and astoundingly detailed sketches. In some instances he had great insight, gained from detailed study of the physics of nature. In yet others, he could not foresee the inherent technical difficulties of his ideas which would take significant changes in technology to achieve realisation. The helicopter and flight are two examples of such ideas. Clearly Leonardo was a clever inventor. Inventors create ideas.

E2.1[2.2] Three of the best-known thought experiments in physics are *Newton's bucket*, *Einstein's elevator* and *Schrödinger's cat*. Search out the descriptions of these famous *gedanke* and explain what they are meant to demonstrate.

2.2 Exercises embedded in the text are denoted with E numbers and projects are denoted with P numbers.

E2.2 One measure of strength of materials used in the paper industry is breaking length. Imagine a sheet of paper of width W, thickness t and length L in the vertical direction suspended from a support at one end. What is the breaking length LB such that the paper sheet will break at or near the support? Is this length influenced by the values of W and t? Draw up a comparative table of breaking length for newsprint, structural steel, *Gladwrap*, titanium alloy, balsa wood and spaghetti.

E2.3 An alternative performance measure for materials is *yield value*. Consider a number of coins (be it a *GBP - Great British Pound*, a *US 25 cent* or an *Australian dollar*) arranged in a single tower carefully balanced, with each successive coin exactly on top of the preceding one. How many coins (what total value of coins) would be required such that the one on the bottom reaches its yield point? What would be the value if each coin were made of milk chocolate?

E2.4 The central post offices of two towns Acton and Boton are 217 km apart. In a charity ride a group of cyclists travel from Acton to Boton and a roughly equal number of cyclists travel from Boton to Acton. Both teams of cyclists commence their journey at their respective post offices. Their maintenance automobile starts at the same time as the cyclists. As the two teams can only afford one maintenance automobile, it is forced to travel back and forth between the two groups of cyclists at a steady 80 km/hr as they approach each other. If the cyclists travel at 31 km/hr, how far must the maintenance automobile travel during the ride?

E2.5 A famous problem in artificial intelligence (AI) is one where a 10 x10 game-board has two corner squares removed as shown in Figure E2.5. The board is to be completely covered with the 2 x1 tiles without overlaps. Can this be done? Offer a proof.

The relation of this problem to AI arises from its feature that one can certainly program a computer to enumerate the various coverings of the board and thereby demonstrate the result, but this is the *brute-force solution*. However, there is a much simpler and more elegant solution, with a compelling proof. Hence in a sense the solution to this problem sits in *AI* as a cautionary tale against *brute-force* solutions to problems. No doubt we will meet similar cautionary experiences in MaT projects.

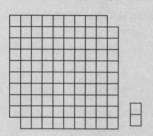

Figure E 2.5 - Modified game board and covering tile

2.1 Inventing, Inventions and Design

Inventions occur when preparation meets opportunity. In the substantial literature on invention and innovation the distinction between these two aspects of generating new ideas has somehow become blurred. There is merit in regarding innovation as a gradual change of an existing paradigm of some artefact or process, whereas inventions are usually associated with some step change in technology, often related to the introduction of a new paradigm for an artefact, material or process. There is also some vague distinction associated with the scale of impact an innovation or an invention might have on the economy.[2.3]

Figure 2.5   The fall of Icarus, by Carlo Saraceni, 16th century Venetian artist Reproduced by kind permission of the Soprintendenza Speciale per il Polo Museale Napoletano - Castel Sant' Elmo

Innovations may be incremental, or step changes or somewhere in between, often associated with relatively shorter-term changes in the technical realisation and exploitation of the invention. Although these distinctions may be economically meaningful, they are less than helpful in assessing the relative creative effort involved in either of these two creative processes. Paraphrasing the words attributed to Thomas Edison, often the invention is the brilliant flash of insight (*the 1% inspiration*) followed by the considerable greater creative effort of the longer term exploitation of the invention through innovation (*the 99% perspiration*).

Often the invention is a result of some recognised need such as the telephone or the wireless (*market pull*) as opposed to the innovation needed for exploiting some new technology such as the computer chip (*technology push*). As an interesting and perhaps instructive example, consider the invention of flight and, eventually its technical realisation, the modern aeroplane. The idea of human flight may be dated to the Minoan myth of Daedalus and Icarus (about 1600 BC), who chose to escape from the then ruler of Crete by constructing wings from bird feathers held together by wax. The mythical story was originally intended as a cautionary tale, suggesting that if we aim too high (flying near the sun), our expectation may well be dashed (Icarus' wings were melted by the sun's heat and he fell to his death).

2.3 See, for example, Mensch (1975); Kondratieff (1984); Radjou *et al.* (2004).

Nevertheless the idea, once born, gave birth to a wide variety of innovative exploitations such as the sketches of Leonardo and the early attempts at flying machines in the late nineteenth century.[2.4] Ultimately it took concentrated effort and deep understanding, resulting from experimentation, between 1899 to 1903 for the Wrights to exploit the idea. In essence then, we see a period of nearly four millennia from idea to realisation. Of course, many other inventions and innovations along the way helped pave the way for the attainment of powered human flight: the internal combustion engine (*power*), steel wire (*control*) and various modern materials such as canvas and aluminium (*aircraft components*). However, none of these other inventions were instrumental in the understanding of the aerodynamics of flight, the real invention necessary for turning and manoeuvring the craft.

In the exploration of inventions and innovations Jewkes, Sawers and Stillerman note the words of Alfred North Whitehead:[2.5]

The greatest invention of the nineteenth century was the invention of the method of invention...The whole change has arisen from the new scientific information...It is a great mistake to think that the bare scientific idea is the required invention...One element in the new method is just the discovery of how to set about bridging the gap between scientific ideas and the ultimate product.

Clearly, we need not be so concerned with the distinctions between invention and innovation, as far as creative energy and effort are concerned. We should focus our interest on the thought experiments required for delivering the realisation of ideas. This process takes place in the world of the engineering designer. As noted by Horst Rittel:[2.6]

It is one of the mysteries of our civilization that the noble and prominent activity of design has found little scholarly attention until recently...Designing is plan-making. Planners, engineers, architects, corporate managers, legislators, educators are (sometimes) designers. They are guided by the ambition to imagine a desirable state of the world, playing through alternative ways in which it might be accomplished, carefully tracing the consequences of contemplated actions. Design takes place in the world of imagination, where one invents and manipulates ideas and concepts instead of the real thing - in order to prepare the real invention.

Engineering designers and technologists develop ideas into realisations. In many situations these two *hats* are worn by the same person. However, while invention may come as a *flash* of genius(*the 1% inspiration part*), the attainment (*the 99% perspiration*) is generally of much longer term. The satisfaction and the wonderful excitement that come with seeing one's idea

---

2.4 Kelly (1996); Chanute (1897); Herring (1897); Crouch (1981).
2.5 Jewkes et al. (1969).
2.6 Rittel (1987).

Table 2.1 Inventions and their technical realisation (Source: Mensch[2.7])

Invention	Date of idea	Technical realisation	Delay (years)
Power generator	1820	1849	29
Electricity production	1708	1800	92
Arc lights	1810	1844	34
Pedal bicycle	1818	1839	21
Crucible steel	1740	1811	71
Locomotives	1769	1824	55
Telegraph	1793	1833	40
Pharmaceutical industries	1771	1827	56
Photography	1727	1838	111
Safety matches	1805	1866	61
Aluminium	1827	1887	60
Refrigeration	1873	1895	23
Dynamite	1844	1867	23
Lead battery	1820	1867	47
Incandescent light bulb	1800	1879	79
Telephone	1854	1881	27
Gasoline motor	1860	1886	26
Nylon	1927	1938	11
Penicillin	1922	1941	19
Polyethylene	1933	1953	20
Radio	1887	1922	55
Television	1907	1936	29
Zipper	1891	1923	32
Ballpoint pen	1888	1938	40
Fluorescent lighting	1852	1934	82
Helicopter	1904	1936	32
Jet engine	1928	1941	13

realised are inevitably delayed. Table 2.1 is a list of some *invention ideas* and their realized innovations. The key feature of the table is the consistent delay between the invention/idea and a final useful product, process or material that seeks to exploit it.

Experienced engineering designers are well used to this delayed gratification. Freshman undergraduates, on the other hand, may find it very frustrating to spend considerable resources on a design problem without the joy of

2.7 Op. cit.

seeing their solutions realised. In addition, this frustration is coupled with the uncertainty of the inexperienced designer about the success or failure of the solution offered. MaT projects, static or dynamic, offer an opportunity for seeing the solution to a design problem not only realized, but also tested, in a competitive environment. If we regard journalism as *literature in a hurry*, clearly, MaT projects are *engineering design in a hurry*.

P2.1 As an example of creative engagement, we regularly invite under-graduates, in their first lecture on engineering design, to construct a type of "bridge", to span the chasm between two supports and to carry only its own weight. The only material available for constructing this bridge is one regulation A4 sheet of cartridge paper. The criterion of performance is the distance spanned.

Most students soon become familiar with this type of MaT exercise. However, we still find a substantial proportion of our classes who find the problem difficult. We always stipulate "simply supported" structures, and the supports offered are most commonly two drinking glasses. Readers are invited to offer solutions to this problem. One solution is given at the end of this chapter.

Although engineering design is regarded, most often by engineers, as the peak of creative engineering activity, its results are often the absence of some effect rather than the presence of some observable feature. Examples are the absence of failures in manufacturing start-ups; the absence of disagreeable

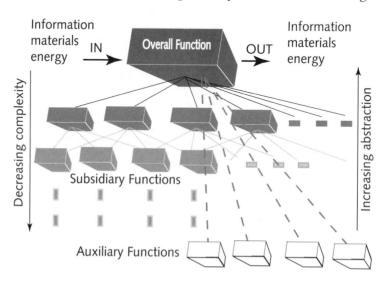

Figure 2.6 Formal abstraction of the generic design problem into its function structure (after Pahl and Beitz, 1996)

behaviour, such as squeaks, leaks, bumps, rattles and unwanted resonances in artefacts, as well as the absence of the need for costly dismantling of equipment during routine maintenance.

Because most of design takes place in the mind, in general, the outcome is limited only by our conceptual wealth. As a direct consequence we need to examine not only the processes of design but also the cognitive development of the designer. By the time we reach for design support tools such as sketchpads, check-lists or computer models, we have reached a partial plan of the style of embodiment to be used. In the early stages of design concept formulation we can gain some help from tools and devices that will reduce the cognitive load on the designer. One often used example of such a conceptual design tool is the interaction matrix[2.8] that permits the relationships between specific design requirements to be drawn out in a way that identifies closely clustered functional connections. The initial step in developing a design concept is to list all the functions the eventual design will need to fulfill. Pahl and Beitz[2.9] specifically identify this formal first step in exploring the nature of the design problem through its function structure. They go on to propose that these function structures, made up of the overall function, its several subfunctions and auxiliary functions, represent the formal abstraction of the design problem. Figure 2.6 is a schematic view of the type of abstract, hierarchical, function structure, proposed by Pahl and Beitz.

Several aspects of this function hierarchy need further elucidation. Firstly, the structure is clearly a hierarchy, made up of many interconnected modules of subsidiary functions and auxiliary functions. In the Pahl and Beitz model, the subsidiary functions directly contribute to the ultimate highest-level overall function (principal design features or requirements), and the auxiliary functions contribute to it in some indirect way (by possibly enhancing, or enabling, one or more of these primary design features or requirements). Auxiliary functions are *value adding* design features that may or may not be articulated in the statement of the primary design requirements; for example, the bottle opener function of a can opener.

Herbert Simon[2.10] argues convincingly about the development of hierarchic systems in a chapter titled "The Architecture of Complexity" in his book *Sciences of the Artificial*. Simon's argument is based on the principle that the stable evolution of any complex system requires sets of stable modules, or subsystems, whose interrelationships, and ultimate interconnection and combination, make up the system in the large. This model will hold equally well for biological, social and technical systems. Simon assigned the term

---

2.8 See, for example, http://www.designmatrix.com/tools/cluster_tools.html; Bisantz *et al.* (2003); Chen and Lin (2002); Agrawal *et al.* (1993).
2.9 Pahl and Beitz (1996).
2.10 Simon (1996).

span of subsystems to the number of distinctly different subsystems from which an overall complex system is composed. The term *span* describes a formal set-theoretical concept for the range (or variety) of subsystems that make up the composite system. As an example, Simon quotes the long chain molecule of polymers, such as nylon, being composed of long chains of identical monomers, that represent large complex systems, with a span of one subsystem.

In order then for a hierarchy to survive, or indeed to evolve, we need the existence, or development, of stable subsystems. This issue brings up an important and inescapable feature of hierarchic representations for complex systems. As the subsystems evolve and eventually combine into the overall system, we must keep track of all the information relating to their interconnections and interactions. As the span (varieties of subsystems) of the overall system increases, so does the information content of these interactions and interconnections. This growth in information content is referred to as increase in entropy resulting from the decomposition of a complex system into its elements (in this connection Simon[2.11] also quotes the work of Setlow and Pollard[2.12] who calculate such increase in entropy in thermodynamic systems).

Finally, another important feature of the hierarchy of functions represented in the Pahl and Beitz model is its relationship to the underlying nature of problem solving and cognitive development of designers. Simon[2.13] identifies two distinct and important types of definitions for problems. One is the *state-definition* of the problem that identifies the way we view a specific design problem now. This is the defined *status quo*, coupled with some implied, or articulated definition of the desired state of the world. For example, *"Too many young children drown in backyard pools"* implies that we would prefer this state of affairs to be somehow rectified by design.

The other is the *process-description* of the problem, or the means by which we hope to achieve a solution state in the future. For example, the statement *"Design and develop safety devices, procedures or laws for the installation and use of backyard pools"* is a *process-description*, or a *map*, for travelling from the current *status quo* of *too many deaths of young children in backyard pools* to a desired future situation of lessening, or perhaps completely eliminating, such accidental drownings.

The function structure is an abstract description of the problem and it is a top-down view of the expectations of the ultimate customers of our design. In a sense, it is related to the *state-definition* of the design problem, as the expression of the designer's ambition for change and improvement in

---

2.11 *op. cit.*
2.12 Setlow and Pollard (1962).
2.13 *op. cit.*

the state of the world. Unfortunately these various formal abstract descriptions of a design problem are of little use in delivering a solution. Often they represent *neat and simple post mortem* reviews of the various functional relationships' specific solutions. Real-life design is rarely neat and never simple. The *process-definition*, or map, for solving the problem usually takes the form of detailed design specifications of the various substructures, and their multitude of interconnections and interactions, that satisfy the functional needs of the overall system. These specifications, or *process-descriptions,* of the design generally result from a bottom-up approach to the design, by considering the simplest (least complex) substructures first and building up the ultimate system from these. This is the problem-solving style embedded in the well-known *quality function deployment* (QFD) procedure.[2.14]

An example of the application of the QFD procedure is provided in the next section. Although, in general, when applying QFD, we need to deal with a known embodiment style of a given artefact, the procedure of identifying the interactions in the *part-deployment* matrix is a direct connection between the *state-definition* (function-structure) and *process-description* (specification) of the design. In this context it is essential to recognise that specification represents the means of achieving the desired functions expressed in the original function-structure.

2.2 An Instructive Example: The Can Opener

If you can talk brilliantly about a problem it can create the consoling illusion that it has been mastered.
Stanley Kubrick, movie director.

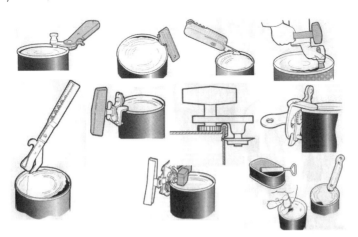

Figure 2.7 Some example sketches of can opener types (After Ledsome [2.15])

2.14 Hauser and Clausing (1988).
2.15 Ledsome (1987).

Table 2.2 Can Opener Design - Desired functional requirements (function-structure)

	FUNCTION	DESIRABILITY 1 = lowest 5 = highest	INFLUENCE ON DESIGN SPECIFICATION
OPERATING FEATURES	Easy to operate	5	Operating loads
	Opens range of shapes and sizes	3	Rectangular cans
	Can be operated by 95% of population	5	Ergonomic design
	Human energy source	3	Ergonomic design
	Other energy source considered	2	Power, current drain
	Portable, compact	4	Size and shape
	Easy to clean	4	Shape, assembly, material surfaces
	Aesthetic appeal	4	Colour, geometry
	Bottle opener	2	Manufacturing detail
	Beverage can piercer	1	Manufacturing detail
	Retains lid	2	Detail design
MANUFACTURING	Easy to manufacture	4	Detail design
	Easy to assemble	3	Detail design
	Mechanically simple	3	Fewer components
	Uses conventional materials	2	Material choice
	Conventional manufacturing technology	2	Manufacturing detail
SAFETY	No sharp edges	5	Detail design, shape and finish
	Cutting/operating mechanism guarded	5	Detail design
	Leaves no sharp edges on can	3	Cutting mechanism
MAINTENANCE	Cutting edge easy to sharpen or replace	3	Material and manufacturing choices
	Operating mechanism parts easy to maintain or replace	3	Detail design
	Opens many cans before maintenance	4	Material and manufacturing choices
MARKETING ISSUES	Cheap to make	3	Complexity, material and manufacturing choices
	Has strong market appeal	3	Appearance and operation
	Brand representation	2	Signage/packaging
	Environmentally safe	5	Material and manufacturing choices

Consider designing a simple device such as a can opener. Its overall function is to open a can. Although this is a relatively well-known artefact, it has considerable design complexity. Some of the questions facing the designer include:

- *Size and shape of cans to be opened* – Cans of circular cylindrical section are somewhat more common than other shapes, but there are some almost rectangular cans with rounded corners(sardine cans, for example), and even completely rectangular cans with sharper corners (fetta cheese containers, for example),

- *Range of can material thicknesses likely to be encountered* – This aspect of the design influences the type of can-piercing method used as well as the forces needed to pierce the can in the first place,

- *Materials and manufacturing processes to be used* – If the opener is to be of limited life and cost to the consumer, corrosion avoidance and mechanical robustness may not be very important subsidiary functions of the design,

- *Energy source* – Is it to be human powered or is it to be power-driven by some other form of energy source?

- *User demography* – Is the can opener to be equally easy to use by both right- and left-handed people (approximately 20% of the population are left-handed), or ageing arthritic people, or indeed people with a wide range of disabilities? These issues speak to the ultimate nature of the artefact to be designed; servicing and maintenance; should the cutting surface(s) and parts or the whole of the driving system be removable for cleaning, lubricating, resharpening and perhaps replacement,

- *User safety* – How are we to deal with avoidance of injury to the user such as sharp or injurious features of the device or ensuring that the device leaves no sharp edges on the can after opening it,

- *Auxiliary features* – Is there a need to capture the can lid? Should we include some other feature in the design (knife or scissor sharpening capability or bottle opener, for example)?

- *Aesthetics* – The appeal of the final artefact to the consumer or client is an important marketing issue. It may be considered that in some cases, apart from the cutting or opening mechanism, the can opener may be completely concealed from view.

Clearly, *"design is a rich study."*[2.16] Figure 2.7 shows some sketches of can openers, possibly as imagined by the designer. The figure indicates some important aspects of the complex function structure of the design problem,

---

2.16 Michael French (1988), in *Invention and Evolution: Design in Nature and Engineering*, section 10.9.3, explores the design specification of a simple clothes-peg and reaches this conclusion.

Krups Open Master - Hand-held electronic can opener with patented bladeless design uncrimps lids with a single-hand operation. Uncrimping action eliminates the danger of leaving cans with sharp, jagged metal edges. Never comes in contact with food so the build-up of grease and food particles which are found on traditional can opener blades is prevented. This bladeless technology eliminates the possibility of tiny metal shavings falling into opened cans. Ergonomically shaped handle for comfortable finger touch operation

Black and Decker Gizmo Spacemaker EM200C - Cordless for flexible use anywhere; swing-out magnet; full charge opens 20 cans; power-pierce cutter for precise cutting; runs around can for "hands free" operation; automatic shut-off; dishwasher-safe cutting assembly; charging base mounts under-the-counter; no-touch disposal of lids

OXO Good Grips locking can opener - Large, comfortable handles, great for people with limited hand strength; contoured handles distribute pressure evenly; snap-button locks firmly to keep can opener in place; Releases with press of a button; oversize knob turns easily. With a lock that keeps the blades tightly in place, the Snap-Lock can opener is the perfect choice for those with limited hand strength or troubled by arthritis. The large curved handles fit into any hand, and the oversize knob turns easily, relieving pressure on the hands. The slightly arc-shaped handles are designed to distribute tension evenly, and the lock can be released by a simple push of the button. The handles are slip-proof, wet or dry, as well as dishwasher-safe

Rösle can opener - Brilliantly designed can opener provides safety and hygiene; cuts along side of can's rim without leaving sharp edge. Opener doesn't contact can's contents; lid can't fall into can; made of 18/10 stainless steel, with rugged plastic head; round, satin-finish handle with hanging ring

Figure 2.8 Some examples of can openers available on the world market. Current prices of these items range between US$ 15 to US$ 50

namely several cutting methods, various can types that might be considered by the designer and a range and style of can opener quality. Figure 2.8 shows examples of can openers available at various sales points, together with their vendor's performance claims.

The designer's task is to assemble the main functions into some ordered list (the *state-definition* of the design) to permit the development of a *map* (set of specifications) for action (the *process-description* for the design). Table 2.2 shows a sample list of design functions for the can opener design.

In Table 2.2 the values of desirability for specific functions are notional and often these values represent the bounds on the eventual design chosen.

In the QFD procedure the information from Table 2.2 is entered in a *product-planning* matrix as shown in Figure 2.9.

In the *product-planning* matrix shown the horizontal list of items represents those specific features of the eventual design that will influence the functional requirements itemised in the vertical list. In some sense, these specific features and their associated functions represent the desired state for the design, or its *state-definition*. The horizontal list of items at the head of the matrix indicates those features of the design that will influence how the desired features and associated functions, listed in the vertical list, will turn out. In an abstract sense, this second list leads directly to the *process-description* of the can-opener design problem. An essential question to ask about the horizontal and vertical listing of design features is *"what input (horizontal list) adjustment need we make, to influence an improvement in the output features (vertical list)?"*

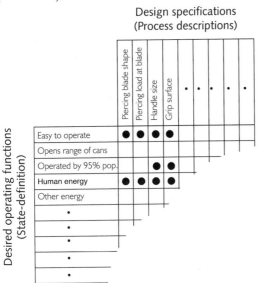

Design specifications
(Process descriptions)

Figure 2.9 Partial "product-planning matrix" for the can opener (from the QFD procedure)

The black dots represent influence values (strong, medium, weak or nil) between these two aspects of the problem statement. One can assign numerical values to these influences, or levels of interactions, ranging from (say) -5 to +5. This range may be chosen arbitrarily since only relative values are of interest. Notably, some influences might be negative, indicating that as we increase some design feature, the influenced functional performance will decrease. For example, as we increase the handle size delivering piercing load to the blade, so the human energy required to pierce the can will decrease. However there is an associated penalty of an increase in size (reduced compactness).

Although the process of listing state-definitions (functional requirements) and process-descriptions (design specifications) appears to be orderly here, in the real development of the design there is a continuous switching between these two aspects of the problem. Lists of functions and design specifications are developed by groups of various stakeholders in the outcome, including various elements of the design community within the

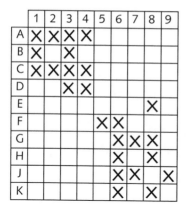
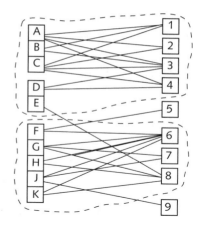

Figure 2.10 A simple interaction matrix and its modified graph

organization. Product design, marketing, manufacturing and production are usually represented in the stakeholder team developing these product-planning matrices. The resulting design specification for the artefact to be built is almost always a result of complex, often informal, negotiation among these team members as they switch their thinking between the two views of the design problem.

The QFD product-planning matrix is just one member of a much larger set of matrix tools used for synthesis in a variety of disciplines. The generic term used for such matrices is *interaction-matrix* or *decision-matrix*.[2.17,2.18] Another useful aspect of such matrix representation of information, especially when used with design interaction information, is the range of computer search methods available for matrix evaluation. Search methods are used for matrix decomposition and cluster analysis. The matrix is replaced by its graph, where the lists of functional requirements and design specifications are nodes of the graph and the filled matrix components are links of the graph.

Both matrix decomposition and cluster analysis examine the graph and identify strongly interconnected subsets of nodes. These subsets are then regarded as independent design problems, weakly connected to other subsets. In this way a large-scale, complex design problem may be reduced to subsets of almost independent, smaller-scale and often less complex, design problems.[2.19] Figure 2.10 is an example of a simple interaction matrix, its graph and a possible decomposition, or cluster structure.

2.17 As for footnote 2.8.

2.18 See also Ng (1991); Kalamaras (1997); Canter (1977).

2.19 Alexander (1964); Alexander and Manheim (1965); Parmee (1997); Bullock et al. (1995); Dogan and Goetschalckx (1999).

In the example of Figure 2.10, the two almost independent clusters of nodes *ABCDE/1234* and *FGHJK/678* appear closely interconnected. Nodes 5 and 9 are *outliers*, and, under certain constraints, may be treated as independent of the other two clusters. The example was designed to illustrate, rather obviously, the clustering of nodes. Currently available decomposition and cluster analysis programmes are quite capable of optimal decomposition and cluster recognition in interaction matrices that have considerably more complex interactions than in the example shown here.

The QFD procedure is also considerably more complex than the one-dimensional presentation offered here. Detailed description of the process is beyond the scope of this book, but interested readers may find useful descriptions in the several references offered in the bibliography.[2.20] This essential tool for reducing the cognitive load of developing process-descriptions from state-definition of the design problem has an important place in early learning about design. The QFD procedure is the first formally and widely used procedure that teaches the importance of relating function structures to design specifications in a clear and unambiguous way. Although it is a descriptive (as opposed to prescriptive – *"You must do it this way, or else!"*) model of early stages of product design, Einstein's view of his models of the physical world should serve as a reminder about the value of such descriptions:[2.21]

> *Einstein regards all physical concepts as free creations of the human mind. Science is a creation of the human mind, a free invention. This freedom is restricted only by our desire to fit the increasing wealth of our experiences better and better into a more and more logically satisfactory scheme. This dramatic struggle for understanding seems to go on forever… The question is not whether Nature is intrinsically relativistic or not. The real question for Einstein was whether or not Nature could be made to look relativistic as expressed through the equations of physics!!*

P2.2 Figure 2.11 shows some examples of hinges used in household joinery. Make a list of the functions for which these hinges were designed. In each case identify the overall function, subsidiary functions and auxiliary functions.

Choose a specific hinge design and draw up a simple table of target design attributes (similar to Table 2.2 for the can opener) for which you think the specific hinge may have been designed.

2.20 Akao (1990); Akao and Mizuno (1994); Terninko (1997); King (1995); Daetz et al. (1995); Cohen (1996).
2.21 Einstein and Infeld (1938).

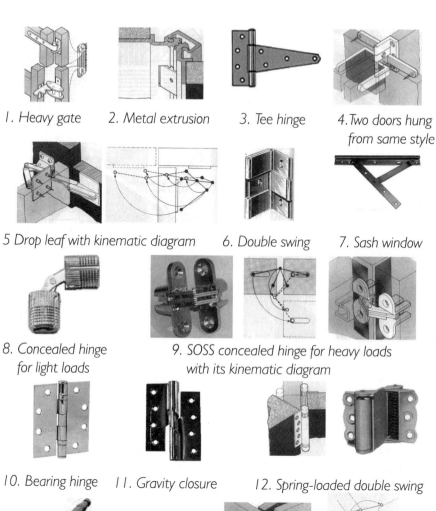

1. Heavy gate   2. Metal extrusion   3. Tee hinge   4. Two doors hung from same style

5 Drop leaf with kinematic diagram   6. Double swing   7. Sash window

8. Concealed hinge for light loads   9. SOSS concealed hinge for heavy loads with its kinematic diagram

10. Bearing hinge   11. Gravity closure   12. Spring-loaded double swing

12. Lightweight cupboard hinge   13. Joinery hinge for more than 90° swing with its kinematic diagram

Figure 2.11 Joinery hinges

2.3 Cognitive Science, Creativity and Analysis in Design

So far in this chapter we have ranged over some broad issues of design, invention and the way in which the personal experiences and mental states (cognitive development) of the designer might influence the outcomes of a design. It is appropriate that we formalise these concepts, inasmuch as they are of some interest to researchers in cognitive science and its application to design. Although Herbert Simon wrote about human decision-making behaviour over half a century ago,[2.22] his views on this topic expressed most cogently in his Nobel Memorial lecture may be summarised by the statement:[2.23]

> It is a vulgar fallacy to suppose that scientific inquiry can not be fundamental if it threatens to become useful, or if it arises in response to problems posed by the everyday world. The real world, in fact, is perhaps the most fertile of all sources of good research questions calling for basic scientific inquiry.

Cognitive science is *"the study of intelligence and intelligent systems, with particular reference to intelligent behaviour as computation."* [2.24] Computer modelling of intelligent (rational) behaviour includes the study of creative problem solving. In one study an artist, Harold Cohen, programmed a robot to produce original works of art.[2.25] The notion of programming and originality may seem self-contradictory. Yet, a MaT project for student designers at Stanford, in the early 1970s, asked for the design and construction of a simple mechanism to produce an original work of art. The best solution in that case employed a small spring-operated mechanical frog, with coloured dye pads attached to its feet. The frog was tethered to a fixed post in the middle of a large sheet of drawing paper and wound up before being released. Due to the uneven surface of the table and the random nature of the wound spring in the frog, as it bounced around its dye-impregnated pads did indeed produce original art. The frog did not behave according to any specific set of rules. It simply exploited its own randomness to perform its task. In this sense, it navigated its (partially constrained) path without maps. In the field of formal representations of knowledge, expert systems provide the procedural rules of navigation, or the operational map. For Cohen's robot, Aaron, the task was to navigate without a map. According to Clancey (1997) this approach is identified as situated cognition:

> Every human thought and action is adapted to the environment, that is, "situated," because what people "perceive," how they "conceive of their activity," and

2.22 Simon (1952).
2.23 Simon (1978).
2.24 Simon and Kaplan (1989).
2.25 Clancey (1997).

what they "physically do" develop together. From this perspective, thinking is a physical skill, like riding a bike. In bicycling, every twist and turn of the steering wheel and every shift in posture are not controlled by manipulation of the physics equations learned in school, but by "recoordination" of previous postures, ways of seeing and motion sequences.

We have considerable empathy with the navigators on ancient ships, with only the starting shoreline to guide them and hopefully seeking a destination, having only vague ideas of where the shoals and sandbars and other hazards might lie along the way to that destination. Of necessity, they had to think their way through their journey and develop their maps along the way. Clearly, the design problem solver has a similar task of navigating the creative challenges offered by the functional definition of the problem, through the hazards offered by various flexible and inflexible design bounds. This too is practical situated cognition.

Creativity is generally regarded as the domain of the artist and designer. In all fairness to writers on the subject of creative behaviour, there are references to creative management,[2.26] and (in the legal sense) creative accounting. Yet, there is also considerable creative energy expended in all the physical sciences at the *coal-face* of research. Typical, well-known examples are the discovery of the structure of DNA, by Francis Crick and James Watson,[2.27] and the proof of Fermat's last theorem, by Andrew Wiles.[2.28] In his commentary on the discovery Wiles wrote:

... suddenly, totally unexpectedly, I had this incredible revelation. It was the most important moment of my working life... it was so indescribably beautiful, it was so simple and so elegant, and I just stared in disbelief for twenty minutes, then during the day I walked round the department. I'd keep coming back to my desk to see it was still there - it was still there.

This then is the looked-for and hoped-for "aha!" experience sought after by all creators of new ideas. It would seem extremely restrictive to limit this experience to the artisans of the world. Psychologists associate creative behaviour with the thought processes of the right hemisphere of the brain. This is in contrast to the actions of the left hemisphere, which is generally associated with more ordered, analytical thought processes. A number of organizations use the Myers–Briggs personality type indicator (MBP) test for assessing creative potential in their staff. This test is based in the Jungian psychology.[2.29] The four personality types and their opposites, identified in the MBP test, are: *extroverted/introverted, sensing/intuitive, thinking/feeling, judging/perceiving*. Although in recent times it has been declared unethical to

2.26 See, for example, Dutta (1991); Wesenberg (1994); Gryskiewicz and Tullar (1995).
2.27 Crick and Watson (1953).
2.28 Wiles (1995); Aczel (1996).
2.29 Jung and Hull (1990).

Table 2.3 Myers–Briggs personality type traits

Extroverted	• Act first, think/reflect later • Feel deprived when cut off from interaction with the outside world • Usually open to and motivated by outside world of people and things • Enjoy wide variety and change in people relationships
Introverted	• Think/reflect first, then act • Regularly require an amount of "private time" to recharge batteries • Motivated internally, mind is sometimes so active it is "closed" to outside world • Prefer one-to-one communication and relationships
Sensing	• Mentally live in present • Being practical and using common sense solutions is automatic • Memory recall is rich in detail of facts and past events • Best at improvising from past experience • Like clear and concrete information; dislike guessing when facts are "fuzzy"
Intuitive	• Mentally live in the future • Using imagination and creating/inventing new possibilities is automatic • Memory recall emphasizes patterns, contexts and connections • Best at improvising from theoretical understanding • Comfortable with ambiguous, fuzzy data and with guessing its meaning
Thinking	• Instinctively search for facts and logic in a decision situation • Focus on tasks and work to be accomplished • Easily able to provide an objective and critical analysis • Accept conflict as a natural, normal part of relationships with people
Feeling	• Instinctively employ personal feelings and impact on people in decision situations • Being sensitive to people needs and reactions is a prime consideration • Naturally seek consensus and popular opinions • Unsettled by conflict; have almost a toxic reaction to disharmony
Judging	• Plan many of the details in advance before moving into action • Focus on task-related action; complete meaningful segments before moving on • Work best and avoid stress when keeping ahead of deadlines • Naturally use targets, dates and standard routines to manage life
Perceiving	• Comfortable moving into action without a plan; plan on-the-go • Like to multitask, have variety, mix work and play • Naturally tolerant of time pressure; work best close to the deadlines • Instinctively avoid commitments which interfere with flexibility, freedom and variety

administer the MBP evaluation unless it is supervised by a professional psychologist, there are a number of Web sites offering the test.[2.30] Table 2.3 gives an overview of the personality traits associated with each personality type.

In spite of its early popularity for assessing creative behaviour, the MBP test is now recognised as a very coarse indicator of potentially creative performance. These types of personality test are, however, widely used to assess motivational behaviour in the classroom and, in some cases, among engineering managers.[2.31]

We could conjecture about the apparently widespread desire to find a means of *a priori* identification of potentially creative capability. In the middle ages, search for the elusive *philosopher's stone,* for converting base metals to gold, was driven by simple greed. But greed cannot be attributed as a driver for Crick and Watson's search for the structure of DNA, nor the search of various mathematics scholars, including Professor Wiles, for the elusive proof of Fermat's last theorem. Perhaps it is the challenge of the hard-to-pin-down dimensions of creative behaviour, or the even more elusive character of creativity, and identifying those who possess creative potential, that drives this overwhelming interest in capturing its essence.

Yet creativity has much more to do with thinking through problems than it has to do with the outcome of creative behaviour. The artwork, the clever design solution or the beautifully prepared chocolate soufflé, are all manifestations of creative behaviour. Everyone is creative sometimes and nobody is creative always. Much more important, to elicit creative behaviour from undergraduates we wish to engender a positive thinking approach to problem solving. In his preface to *Teaching Thinking,* Edward de Bono wrote:[2.32]

> *Thinking is the most awkward subject to handle. It always involves resentment. It is felt that you are suggesting that the thinking of other people is not as good as it might be - or, worse, that your own thinking is better...The difficulty is that thinking is so closely involved with the ego that in all except young children thinking is the ego. Criticize someone's thinking or suggest an inadequacy and you threaten that person's ego in the same manner. Very few people can so detach themselves that they can look at their own thinking on some matter and describe it as feeble.*

Most academics recognise the common behaviour when students are expected to respond to a new problem in the short term (verbally in class or on a quiz). Avoidance, deferral to authority or cheating are common experiences in such a situation. In cognitive therapy, this phenomenon is referred to as coping.[2.33] When we ask anyone to solve a puzzle in public,

2.30 URL: www.humanmetrics.com/cgi-win/JTypes1.htm; haleonline.com/psychtest.
2.31 Fodor and Carver (2000); Fodor and Roffe-Steinrotter (1998).
2.32 de Bono (1978).
2.33 D'Zurilla and Chang (1995); Amirkhan (1994).

Figure 2.12 H1 – Harrison's first maritime chronometer (Photo by Author). H1 was originally housed in a cubic case 1.22 m (4 feet) in all three dimensions and weighed 33.5 kg (75 lb). H4 is 125 mm (5 inch) diameter and weighs 1.3 kg (3 lb) (after Sobel, 1998)

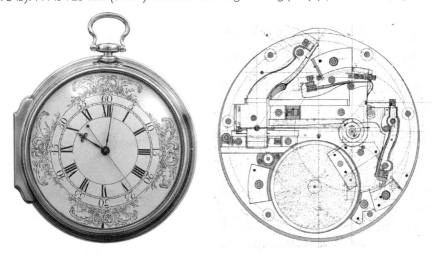

Figure 2.13 H4 – Harrison's final prize-winning maritime chronometer (© National Maritime Museum, London) and construction detail for H4. The drawing of the layout of the movement of the H4 timekeeper by John Harrison (1693-1776), c.1759 (pen and ink on paper),WCC106112, reproduced here with permission of The Worshipful Company of Clockmakers' Collection, UK / Bridgeman Art Library

similar coping reactions often follow. But great inventors of the past were not merely coping with a problem, they were engaged with it. The ultimate key to eliciting creative behaviour is to find means of engaging the problem solver with the problem. Once engaged, often almost obsessive commitment follows (should this be termed "being married" to the problem?).[2.34]

A fascinating example of engaged, almost obsessive, problem-solving behaviour is chronicled by Dava Sobel in her book *Longitude*, about John Harrison, the eighteenth century inventor of the marine chronometer.[2.35] , A meticulous, almost reclusive, craftsman with no formal schooling, Harrison spent most of his adult life in the construction of the chronometer. The Longitude Act of 1714 decreed a prize of £20,000 for a *"method to determine longitude to an accuracy of half a degree of a great circle."* When ships went to sea, in the absence of other navigational aids, the timekeeper was the most reliable means of establishing longitude. When the Longitude Act was proclaimed, the best timekeepers would lose or gain as much as half an hour a day. Harrison completed his first marine timekeeper, named *H1*, in 1735 after five years of work, but he was dissatisfied with its performance, even though it was only errant by no more than a few seconds per day. His prize-winning watch, *H4*, was completed in 1759, nearly thirty years after he started working on the design, and was accurate to within 1/3 of a second per day. Inasmuch as his family did not receive the Longitude prize in full until after his death, it is unlikely that he was driven by greed. Clearly, he was completely engaged with his problem. Figure 2.12 is a photo of *H1* (Harrison's first chronometer) and Figure 2.13 is that of *H4*, the prize-winning marine chronometer completed in 1755; both are housed in the Royal Observatory at Greenwich.

It may be conjectured (provocatively) that searching for an *a-priori* detection of creative potential is, without a suitable problem, akin to trying to detect swimming potential without a swimming pool, or tennis potential without a racquet and ball. In spite of that there is great value to be found in the use of creativity enhancement tools. In *Conceptual Blockbusting*, Adams[2.36] explores several sources of negative influences (conceptual blocks) on creative problem solving. Once we become aware of these conceptual blocks we are able to cope with them in problem-solving situations. Several references emphasise the close connection between visual perception and creative behaviour. Enhancing of the visual character of problem solving can substantially influence the creative outcome.

2.34 Perry et al.(2002); Wentzel and Watkins (2002); Campbell (2001); Bruning and Horn (2000); Reiter-Palmon et al. (1997).
2.35 Sobel (1995).
2.36 Adams (1974); McKim (1972).

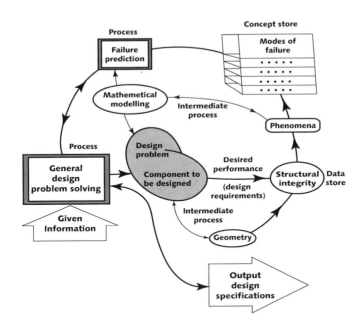

Figure 2.14 Concept map of a generic design problem (after Samuel and Weir, 1995)

Cognitive maps, or mind maps as well as other visual representations of knowledge, can substantially influence the exploration of a problem.[2.37] In essence, however, having a problem worthy of exploration and proper engagement with that problem are the real keys to creative behaviour and creative outcomes. As suggested earlier in this chapter, inventions (*creative outcomes*) happen when preparation (*engagement*) meets opportunity (*a worthy problem*).

Figure 2.14 shows an example of a concept map (*mind map, cognitive map*) for the engineering design of generic engineering components. The diagram shows the concepts (ideas) involved and the interconnectedness of these ideas in solving the problem. Notably, concept maps are very much personal representations of knowledge. In general, one person's concept map may look quite different from someone else's. They are essentially means of communication with oneself for the purpose of clarification of ideas.

E2.6 What is the clock accuracy required at the equator to detect location within half a degree of a great circle? What distance does this represent? Given that current clock accuracy (Cesium atomic clocks) is of the order of 1 second in 25 million years, and that we have decided to

2.37 Samuel and Weir (1995); Terzis (2001); Alexander (1991); Parrott and Strongman (1985).

use such a clock, within what accuracy can we determine location of an object on the Earth's surface?

E2.7 Draw up a mind-map for the design of one of the hinges shown in Figure 2.11. The nodes of this map should be the various functions that the chosen hinge needs to fulfill.

2.4 Engagement and Learning

In the Socratic Dialogue,[2.38] *The Meno*, Plato seeks to explore the notion of learning through the dialogue of Socrates with the fifth century Athenian sophist Meno. The term "sophist", used by Plato, has the following synonyms in the thesaurus: doctrinaire, obfuscator, obscurantist, nit-picker, scholar, theoretician. One could charitably assume that Plato's intention was the term *scholar*, but we should not be alien to the notion that he was well aware of the other meanings of the term. In the dialogue:

Socrates then argues that all so-called learning is in fact the recovery of pre-existent knowledge in the soul, and if virtue is teachable it must be knowledge.

Although this argument relates to the ambiguous concept of virtue, Socrates also goes on to explore knowledge of a more concrete nature. In the presence of Meno, he asks a slave boy to construct two squares, with the second square having twice the area of the first. At the beginning the boy seems to be confident of his knowledge of geometric construction. However, as the dialogue goes on he becomes less certain and in the end quite confused by the concept of doubling the area of the square.

Socrates: Observe, Meno, the stage he has reached on the path of recollection. At the beginning he did not know the side of the square of eight feet. Nor indeed does he know it now, but then he thought he knew it and answered boldly, as was appropriate – he felt no perplexity. Now however he does feel perplexed. Not only does he not know the answer; he doesn't even think he knows.

Meno: Quite true.

Socrates: Isn't he in a better position now in relation to what he didn't know?

Meno: I admit that too.

Socrates: So in perplexing him and numbing him like the sting-ray, have we done him any harm?

Meno: I think not.

Socrates: In fact we have helped him to some extent towards finding out the right answer, for now not only is he ignorant of it but he will be quite glad to look for it. Up to now, he thought he could speak well and fluently, on many occasions and

---

2.38 Plato and Guthrie (1956).

before large audiences, on the subject of a square double the size of a given square, maintaining that it must have a side of double the length.

Meno: No doubt.

Socrates: Do you suppose then that he would have attempted to look for, or learn, what he thought he knew (though he did not), before he was thrown into perplexity, became aware of his ignorance, and felt a desire to know?

Although pedagogues are well aware of how the Socratic dialogue approach, of merely questioning students, helps to guide them to a discovery or the seeking of knowledge, we are also keenly aware of how Socrates died at the hands of the Athenians he sought to influence with his clear philosophy. As professional educators, we are placed in a very tricky position with respect to our students and their learning processes. We could continue to make them aware of their ignorance, in the hope of eliciting some enthusiasm or engagement for learning to overcome it. Yet in some students we could expect the Athenian response of anger and frustration for being exposed, even to ourselves, as more ignorant of some issues, than we had been previously aware. Asking, or perhaps even demanding, a group of designers to think on and respond to an unfamiliar problem, even in the form of the relatively passive Socratic dialogue, can elicit a serious conflict situation.[2.39] We, the educators, may be already engaged with the problem, but the student needs to be guided gently into the miracle of engagement, without eliciting a counterproductive coping reaction.

Jean Piaget,[2.40] the pioneering Swiss philosopher and psychologist, wrote about the development of thought processes in children. He proposed the following major stages of cognitive development:

Sensimotor stage(birth to 2 years): As the name suggests the acquisition of knowledge is through sensory inputs.

Preoperational stage (2 to 7 years): The most important stage of development of thinking during this period is the ability to represent objects by symbols and recognition that a thing or object may be described without it being necessarily present. This stage of cognitive development also includes the process of conservation of mass volume and size even when things change shape.

Concrete operational stage (7 to 12 years): During this stage the most significant cognitive element to develop is grouping. This is the capacity to group objects that are alike (have strong generic similarities), and to understand the notions of number, space and classification of objects.

2.39 Likert and Likert (1978); Falk (1982).
2.40 Piaget and Duckworth (1970); Gruber and Voneche (1977); Boden (1979); Vidal (1994).

Formal operational stage (12 to 15 years): Capacity to handle abstraction and manipulation of abstract ideas.

When a student is placed in a cognitive conflict or threat situation, such as exposure of ignorance or in a formal, time-limited examination, there is some evidence of regression from the formal operational stage of thinking to the concrete operational stage. In such situations the student will attempt to identify something familiar in the problem and explore that rather than seeking to manipulate the abstract ideas involved in the problem statement. This *threat response* has been noted by educators informally, but not explored experimentally until recently.[2.41, 2.42] As an alternative learning approach to the Socratic dialogue and its associated coping response, we might try to deal with established problems where the answers and responses are already known. We could then guide the student through the problem and solution and point out the important steps of reasoning along the way. In general, this is the approach taken in many of the engineering sciences. An unresolved issue with this alternative approach is the transportability of experience from the familiar to the unfamiliar, or from the *established* to the *emergent* situation. The essence of this difficulty may be encapsulated in the aphorism: *One can continue to polish the bullock cart but never get to the aeroplane.* McKim in his book *Visual Thinking,*[2.43] encourages students of design to *"make the familiar strange and the strange familiar."*[2.44]

One approach in engineering design learning is to expose students to the excitement of success in solving an unfamiliar design problem. This process also elicits some level of engagement not normally encountered in the early stages of learning in the classical sciences. The notion of moving from success to engagement and learning is not a new one. Both the Montessori method of childhood learning and the Yamaha music teaching method[2.45] concentrate on early success within the medium of learning encountered. This type of student centering is the essence of early engagement[2.46]. Finally it is worth noting that even in freshman-level natural sciences the issue of engagement is a difficult one to address. Epstein[2.47] suggests one tactic in this context:

The class is conducted only through student questioning. There is virtually no lecturing as such, and the dialectic tone is maintained by the rough rule that instruc-

---

2.41 Samuel (1984).
2.42 Field *et al.* (2003).
2.43 *op cit.*
2.44 See also De Mille (1994).
2.45 Montessori(1976); Lillard (1996, 1997); Vaughn and Stairs (2000); Korfmacher and Spicer (2002); Cox and Rowlands (2000); Miranda (2000); Byo (1993).
2.46 Samuel (1986).
2.47 Epstein (1970).

tors are not to be talking more than half the time. …We assert that the natural curiosity and motivation of students will effect participation…

In the preceding discussion we have ranged through a variety of learning techniques that may be used for engaging student interest. This need *per se* begs the educational policy question of why this arousal is indeed necessary, particularly in vocationally oriented courses like engineering. Models of the tertiary educational process regard it as a social enterprise within which the student and teacher become partners in learning. It is an unfortunate result of substantial resource cuts to universities in the last few decades, that in the "corporate university" of today, education has become a burgeoning business and in many cases the graduating student is seen by academic administrators as a "product", with quantity often exceeding quality in importance. This is a condition entirely due to the political zeitgeist that market forces should determine the quality of tertiary education. Survival of the fittest may be an approach well suited to commerce, but certainly not to the more delicately structured educational system. It is still the serious educator's responsibility to ensure that engagement and learning takes place, even in the face of almost unsustainable student–staff ratios.[2.48]

In engineering courses (and in many other vocationally oriented courses) learning requires several alternate modes of thinking. Applying only gross generalities, we can broadly identify two streams of learning in engineering. One stream (most of the introductory engineering sciences) uses analytical problem solving, associated with a serial-linear thinking approach. Problem solving in this stream deals with *established* situations. The other stream of learning, including engineering design, is associated with engineering synthesis, where the problems address *emergent* situations. The thinking required in synthesis is diverse, requiring informal, creative, inventive, visualising (occasionally radical, way-out, *yellow-hat*) thinking, as well as an organised, judgmental, evaluative decision making (*black-hat*) style of thinking. During their late school years, preceding university entry, students become used to the more formal analytical style of thinking to which they are exposed. It may be conjectured that the creative, explorative approach to problem-slowing, practiced by young children when first exploring their world, has become withered (*"Use it or lose it"*) in the formal, almost regimented, learning programmes of the late school years.

When first introduced to engineering synthesis in design, the weakness of their creative thinking style becomes exposed. Often avoidance, deferral to authority (*"Let's look it up in a book"*) or cheating (copying of others) takes place. MaT projects provide an early opportunity in engineering to engage student interest in synthesis without the associated fear of exposure of

---

2.48 See, for example, Moore (2005).

weakness in thinking styles. These projects may appear (to the uninitiated student) as *play-exercises* in design. With these projects one is not exposed to the serious consequences of failure in "real" design. It is clear from the outset that MaT projects are free from any serious risk of exposure to great economic losses or damage to personal prestige. Removing these real-life design pressures from a design problem (still an *emergent* situation) can place substantially greater emphasis on the creative, synthetic elements of problem solving. Careful organisation, follow-up discussions and evaluation can help to draw out the more serious learning values in the work.

2.5 Some Examples of Generic MaT Projects in Engineering Design

In this section some non specific MaT projects are described. They can form the basis for designing and planning a wide range of MaT projects with specific objectives. The projects may be used to engage student's minds while they are still unfettered by the need to be highly analytical. In all cases the need to think through the problem is stressed, but the usual time constraints imposed on these short-term projects generally prohibit detailed or extensive analysis.

P 2.3 Projects Based on Gravitational Potential Energy

- Transfer energy from a golf ball, tennis ball, or any other mass, to drive a vehicle over some convenient test course.

- Use one form of potential energy (fluid- or solid particle-based, such as water or sand raised to some height) to do work (*e.g.*, to raise a mass).

- Transfer potential energy of a mass to generate a suitable impact to launch a vehicle or to destroy some structure designed to protect a valued object (such as a fresh egg or a delicate plaster article).

Some examples are:

- Design a vehicle, using only the energy available from a standard golf ball, falling under the action of gravity from a vertical height of two meters, to carry a payload of 0.2 kg. – *Criterion of performance* is the distance travelled on a relatively smooth horizontal surface.

- Devise a simple track to permit a coin (The US *quarter* or the Australian *dollar* are suitable coins, the UK *pound* coin is not a regular circular coin) to roll along a smooth floor for some distance. – *Criterion of performance* is distance travelled. In case of a tie the criterion will become speed over a designed test distance.

- Design and build a structure that will protect a fresh egg under it when a standard house brick is dropped on it from a vertical height of 1 m above the structure. You may use only newsprint and adhesive tape as materials of construction. – *Criterion of performance* for all successful structures is the mass of the structure.

P 2.4 Projects Based on the Use of Elastic Energy

- Mechanical springs of virtually limitless variety are available to use as the drive for vehicles, or to launch objects over some distance in air or water.
- Elastic energy is available from pneumatic devices such as balloons or air compressed into some container (*e.g.*, standard 1 litre plastic beverage container). This energy may be used as a prime mover. However, with compressed gases (air included) special care is needed to ensure safety during the project.

Some examples are:

- Design and construct a simple vehicle to be driven over a generally horizontal test track using a standard mousetrap as the prime mover. The track surface is constructed of brick paving so the vehicle may experience some bumps and rolling friction. – *Criterion of performance* is distance travelled.

- Using only the energy available from three standard rubber bands (supplied) design a means of launching a mass (two standard AA size batteries) over some distance. – *Criterion of performance* is horizontal distance covered by launch vehicle as measured from launch pad.

- Using the energy available from a standard 1 litre plastic beverage container (Coca Cola or lemonade bottle), when pumped up with a hand bicycle pump to a maximum pressure available, construct some means for raising a mass over 2 m vertical height in minimum time (*i.e.*, rate of working, or power output). – *Criterion of performance* is power output in watts.

P 2.5 Projects Using Thermal and Electrical Energy

- Thermal energy is readily (and safely) available from candles, hot water, or small quantities of burning oil. Hot water heating elements offer yet another source of such energy. The design challenge is to harness and transform this energy to do some observable and measurable work in a competitive environment. Thermal energy is also available from direct heating by electrical resistance devices such as *Pyrotenax* cables[2.49] or wound wire electric resistors.

- Portable electrical energy is available from batteries or solar cells. In general, dry cell batteries are the preferred storage medium in MaT projects, because they are readily available and are relatively inexpensive. Electrical to mechanical energy converters (electric motors) are also readily available from hobby shops and hobby suppliers. If this type of energy source is to be used in the MaT project then batteries and motors should be supplied, or carefully specified to ensure a *level playing field* for all participants.

Some examples are:

- Design and construct a device that uses the energy available from the burning of a single candle to raise a mass, m, to some vertical height, h. – *Criterion of performance* is the work done ($m \times g \times h$).

- Given a small electric (3.5 V motor) and two standard AA size batteries, design and construct a vehicle to transport a small payload over a test track in minimum time. The test track is the long dimension (50 m) of a shallow reflection pool 200 mm deep and 2 m x 50 m in plan. Initially all vehicles able to cover the 50 m will be compared in performance by a time trial. – *Criterion of performance* is the minimum time taken to cover the 50 m distance, or the maximum average velocity over the measured test track.

These few examples serve to illustrate the range and variety of MaT projects available to anyone willing to use radical design thinking. In general the problem should be simple and easy to understand. The challenge should be in the comparison of specified performance criteria to focus the minds of participants on the engineering science of the project, rather than on constructional detail. However, if there is evidence of ingenuity or constructional elegance in the design this aspect of the project, although not of primary importance in meeting performance criteria, should also attract some extra reward.

As a further resource for projects Table 2.4 provides a brief list of attributes for MaT project selection. One attribute of such projects is the basic human need (food, shelter, transportation, environmental control, recreation, various tools and defence systems) with which it is associated.

2.49 Pyrotenax is the now almost generic trade name for a range of externally insulated electric heating conduits, enclosed in a range of sheaths including copper and steel. These cables are the basis of under-floor heating elements and electric oven elements.

Table 2.4 A selective MaT project attribute list

Project type (and associated human need)	Output performance goals
(Food or food preparation) – kitchen pantry or supermarkets shelf stacking/handling	
Payload-carrying vehicles that can turn corners or move about in restricted spaces	Manoeuverability and guidance
Food container or delivery devices. e.g., "egg drop," Devices for handling and sorting. e.g., pick up, delivery and size sorting of (say) two different sizes of glass marbles, or golf balls and tennis balls	Safe delivery and/or management of cargo
(Shelter) – Housing and protection. e.g., protection of egg from a falling brick using a paper structure	Structural performance under steady or impact loading
(Transportation) – Delivery of payload; water, land and airborne vehicles; unusual delivery systems. e.g., delivery of payload by climbing a vertical rope or wall	Speed and payload to tare weight ratio; guidance and environmental control
(Environmental control) – Vibration and impact resistant structures; energy conversion devices. e.g., spaghetti tower to support egg; use heat from candle to raise a mass;	Light vibration resistant structures; conversion efficiency
(Recreation) – Sport, arts and hobby related devices. e.g., ping-pong ball thrower, tennis ball catcher; creation of art by a mechanism; airborne endurance device; coil spring or elastic driven musical scale generator, walking on water	Time trial, performance demonstration, to "show it works"
(Tools and defence systems) – Surveillance, motion and vibration detection devices, simple tools used with paper or soft materials. e.g., thermal paper punch; simple mechanical vibration sensor	Demonstration; simplicity or operational elegance

2.6 The Benefits and Costs of MaT Projects

Creativity and eclectic knowledge of engineering sciences may help in solving problems of an analytical nature, but for synthesis there is a need for open-minded engagement with the problem. Experiences with analogous problems can provide valuable insight and a significant reduction in the search for possible solutions. Jacob Rabinow again:

> Shockley surmised that creativity must not have a simple functional relationship to working with the brain; if it did, it would be more or less equal among most people. He postulated that if an invention requires the combination of four ideas, and a man can put the four together, he may get it. A man who can only put three ideas

together will never get it. And so ...he showed that if a man can put together twelve ideas, and a second can do only six he's not just twice as good as the second fellow; he's some hundred or thousand times better. Consequently Shockley concluded that small differences in idea-associating abilities of individuals make a tremendous difference in their creativeness.[2.50]

Many engineering schools use MaT projects as an introduction to design. Although these projects are fun for students and teachers alike, their value as an educational tool has not been seriously questioned in terms of educational research. Economic pressures on universities have signalled a need to examine the educational effectiveness of every aspect of what we as educators do with our resources.

It is a generally accepted (though not always verified) feature of academic course planning that continuous quality assurance of a course mix will identify, and occasionally weed out, those aspects of a course that make less than effective use of academic resources. The times allocated to the engineering sciences, accepted as the basic "building blocks" of engineering courses, represent the starting point in many cases for course planning. In many engineering schools design is still regarded as a key part of "bedding down" the basic analytical approaches to which students are exposed in the engineering sciences. However, design is a significantly labour-intensive programme, requiring close mentoring by trained staff. Consequently, it is quite appropriate to question the learning values imparted by participation in MaT projects. Most educators, who use questionnaires to evaluate student responses to these projects, report positive symptoms.[2.51] *"Students enjoy the MaT projects,"* or *"Participants feel they have learnt something practical,"* are common responses to questionnaires. Yet rarely do we see evaluations that show some degree of self-awareness.[2.52] Moreover, when students are questioned about the value of their own educational programmes they are being asked to respond about educational science matters we should not expect them to have understood or even recognised.

Nationwide ingenuity competitions[2.53] are only concerned with intellectual engagement and they need not be overly troubled with educational issues. However, where MaT projects are used in a formal educational stream, where they take up valuable time, replacing other forms of learning programmes, their effectiveness as an educational tool deserves to be

2.50 Jacob Rabinow, *The process of invention* in De Simone (1968). In this context Rabinow was referring to William Shockley, *On the statistics of individual variations of productivity in research laboratories,* Proceedings of the IRE, March 1957, pp. 279–290.

2.51 Several references dealing with this aspect of MaT projects are noted in Chapter 1. "Introduction".

2.52 Samuel and Lewis (1974) found a "castor-oil" reaction of some students to their experiences in a project suggesting the response "I did not enjoy it, but it was good for me"; see also White and Frederiksen (1998); Gladman and Lancaster (2003); Ross et al. (2000).

2.53 www.usq.edu.au/warman_comp.

addressed. The following list of issues requires careful examination when setting and delivering MaT projects as a means of encapsulated design experiences.

- Students enjoy MaT projects for different reasons from those of their teachers. They can be more entertaining than formal exercises in design, particularly if they are permitted to proceed without any analysis or design planning. At first sight, all that seems to be required is that the constructed device should work, albeit desirably better than its rivals. Treated in this manner, MaT projects could become undisciplined play-exercises in design. It was perhaps this view that made some older established universities wary of their regular use in course work.

- In well-organized MaT projects, early formal design planning is needed to generate the solution. This aspect of MaT projects needs careful guidance through mentoring, tutorials and formal reporting sessions. In general, students should be required to submit logs of their sketches and ideas as they progress through to their eventual solutions. These idea-logs represent snapshots of the student's thinking processes during solution.[2.54] Some examples of this type of idea-logs are presented in Chapter 1, "Introduction". Other examples are included with the case examples presented in Chapters 5 and 7.

- In general, MaT projects are group exercises in design and often there are several group dynamics issues that influence the way these projects proceed. Care and guidance are needed to ensure smooth operation of design groups in their progress towards team behaviour with shared goals. They should be interviewed as individual teams during the mentoring sessions to ensure they retain the privacy of their particular approach to solving the design problem.

- Materials and project components should be chosen to ensure the cost of these items does not impose an unnecessary burden on the department or institution offering the project. All materials and components should be available either from toy or hobbyist's stores or local hardware suppliers. Moreover, the total cost of MaT projects should be limited to some nominal value (*e.g.*, student teams should be prepared to sell their device for – say – A$10, US$7 or 6 Euro).

- The manufacturing skills required for constructing project components should be such that students with no access to specialist workshop equipment should not be disadvantaged. Institution workshops

---

2.54 Samuel and Weir (1999).

should be expressly forbidden territory for these types of projects. This essential "levelling of the constructional playing field" is a serious requirement for well-planned MaT projects.

- Student teams should be expected to submit a report of their project, including plans and performance estimates. As well, there is a need to establish some form of personal contributions by team members. The simplest method of achieving this is to ask for a detailed design project diary to be submitted with the report by each team member.

- Teams should be asked to *sell* their solution idea to a *learned audience* of their peers in an informal *sales presentation*. This presentation may be held over until after the head-to-head testing of devices, to ensure some degree of security and protection of ideas.

- Some students might feel threatened by the need to submit a *working prototype*. This concern might be a function of personal background or individual cultural development. Consequently, success of the working prototype should represent only a relatively small percentage of the value of the total project. The remainder of the project value may be allocated to reporting, sketch plans, performance estimation, presentation of ideas to peers and running performance. In this way the threat of constructional difficulties or performance concerns may be reduced substantially.

- Clearly, these projects are labour intensive for both staff and students. However, when properly resourced and carefully managed, the substantial learning values of MaT projects may be identified as follows.

  — *Engagement* – This is perhaps the most significant outcome as it results in emotional involvement with the solution of challenging problems.

  — *Conceptual thinking* – This embodies the early recognition that for a useful design outcome a useful design concept is needed in the first place.

  — *Attention to detail* – The value of this aspect is a hard-learnt experience occasionally at the cost of many experimental failures. Nonetheless, it is often exemplified by the success of carefully designed and constructed successes.

  — *Experience with design synthesis* – The MaT project is probably the first introduction of design synthesis to which students are exposed.

— *Associated issue*: These projects are designed to elicit a great deal of interest in synthesising a solution to a new problem. This can only occur if the projects have a deal of freshness about them. As well, they should be learning experiences that both mentors and students go through together for the first time. Consequently, annual *repeat* MaT projects should be avoided.

— *Team problem solving* – Tertiary institutions tend to encourage and reward individual performance. MaT projects provide, often a first time, experience of participation and leadership in team problem solving. The value of this experience cannot be overstated. An essential part of problem solving in groups is the bonding of the group into a team with shared goals and ideas.

— *Success and self-belief* – Competing and even winning in head-to-head competitions in design can be very stimulating to participants. Seeing your own ideas come to fruition in the short-term is a lasting experience in design projects.

— *Associated issue* – Growth in self-confidence can make a MaT project very enjoyable. The converse (losing and failure to perform) can also leave a lasting and perhaps negative impression on a participant. Projects should be planned, as far as possible, to avoid this type of experience.

• *Care in specifying MaT projects* – As already noted earlier, when we expect students to think creatively in a problem-solving situation, unless properly engaged with the problem, some will find avoidance, deferral to authority or cheating as a means of coping with the problem.

— An example of avoidance is seen in cases where a student team might plan to exploit design specifications to advantage. Virtually all design specifications contain *loopholes* that permit *bypass solutions*, that effectively avoid and bypass the real spirit of the problem-solving exercise. In most formal design experience these types of solutions are expressly disallowed and perhaps even penalised. In MaT projects ingenuity and creative solutions are encouraged.

— Early judgement of ideas is expressly discouraged. Careful examination of the design specifications and design constraints can lead to bypass solutions that may significantly reduce the work involved in constructing a "real" solution to the problem. For example, in a design project calling for vehicles that could transport a large payload while "airborne" for a substantial period of time, one solution offered was a large piece of dry ice (solidified carbon-dioxide). It may seem appropriate for service providers (such as professional designers) that they should indeed exploit the specification of a design problem to their own advantage. It must be made clear at the outset of a MaT project that this is only acceptable as long as the resulting design is within the spirit of the exercise.

– *Associated issue* – MaT projects are creative engagement projects and one of the most engaging aspects of them is the opportunity for unfettered creative problem solving. In giving free rein to creative synthesis we can expect solutions that do not quite fit into the expected design mould planned for the project. Typical examples are: rocket-launched devices when gas jet launching was specified or gas jet propulsion rather than spring or elastic propulsion when strain energy was specified as the drive source. In these cases performances may be divided into separate categories. However we deal with such situations, it is essential to maintain acceptance and proper recognition of creative options.

• Finally, we need to consider the cost of educational resources spent on MaT projects. If introduced early in design programmes and if they are designed to represent one of many other, perhaps more formal, experiences in design, their benefits certainly far exceed the costs of resources spent on them. In the context of the overall engineering course the MaT project can add memorable spice to an otherwise necessarily constrained diet of formal educational experiences.

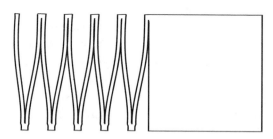

Figure 2.15 A4 paper bridge solution to project P2.1; The figure shows a partially sliced up page of paper. The ends need to be attached to the supports

Figure 2.15 shows one of many possible solutions to the A4 "paper bridge" project offered at the beginning of this chapter (P2.1). The sheet may be cut as indicated and the ends tied to two drinking glasses. This, of course, violates the requirement that the "bridge" should be "simply-supported", but it also illustrates the "flexible" nature of engineering design requirements. The length of the "bridge" is only limited by one's skill with an Exacto knife.

P2.6 *The student paper competition.* An interesting and challenging problem is one that leads to the early development of design analysis, testing and reporting skills. Perhaps the most effective approach for engaging student interest is to offer them the opportunity to explore a problem that falls well within their ambit of interests. This type of problem can serve as a useful first introduction to MaT projects. Typical examples of personal interest are:

– *Sport* – Tennis rackets, ski poles, bicycle gear, surf boards, canoe paddles and rock climbing equipment.

– *Recreational* – Deck chair, clothes line, beach umbrella, fold-up rain shelter, fishing equipment, spinning tops.

– *Household* – Bottle and can openers, waste management systems, cleaning equipment, gardening equipment.

In each case the objective should be to identify strong and weak features of design detail and to explore design alternatives. Wherever possible students should be encouraged to suggest and design simple testing equipment to test the benefits of any improvement proposed to existing equipment. The intended outcome of this project is a report (the student paper).

P2.7 *Exploring simple dynamics* – In this project a simple dynamical model is required for identifying the behaviour of some well known device.

(a) *The spinning pin* – When a simple drawing pin is held by its stem and spun between thumb and forefinger and then released onto a smooth flat surface it will spin for some time, akin to a spinning top. Figure P2.7a shows the spinning pin schematically. This project is intended to

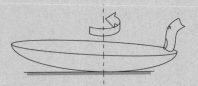

Figure P2.7a. The spinning pin   *Figure P2.7b. Spinning celt or wobbly stone*

explore experimentally how the pin design may be altered to improve its spin behaviour. *Criteria of performance* are spin stability and spin time. A well designed pin should regularly spin for between 60 to 120 seconds.

(b) *Celts*[2.55] – This is an interesting and puzzling dynamical device (refer to Figure P2.7b), also called a "wobblestone." When the celt is spun on a flat surface, it has a tendency to prefer one direction of spin. If the celt is spun in the non-preferred direction it slows down and reverses its direction of spin. Moreover, it may be given a rotary motion by simply depressing one end then releasing it. This project is intended to explore the effect of geometry on the spin behaviour of the celt. *Criterion of performance* is the angle of reverse spin diven by the instability generated in forward spin.

2.7 Chapter Summary

Myself when young did eagerly frequent
Doctor and Saint, and heard great argument
About it and about: but evermore
Came out the same door where in I went.
Omar Khayyám

This chapter presents a coherent theory of engineering design and some philosophical views about invention, creativity and learning. The intention of the discussion and examples offered is to elicit engagement and thinking about the nature and practice of engineering design problem solving.

Major themes explored in this chapter:

- The two identities of design problems, mapping from *state-definition* of a problem to the *process-description*, or abstract design specifications;

- *Interaction matrices* that permit search for clusters of strongly interacting elements of the design;

- *Engagement with the problem* as a major motivator for creative problem -solving behaviour;

- A brief compendium of generic MaT projects in engineering design;

- *Benefits and costs* of MaT projects and the formal planning for MaT projects in design.

2.55 Blackowiak, Rand and Kaplan (1997); Walker (1979).

3
THE GENESIS AND DEVELOPMENT OF MaT PROJECTS

Science is organized knowledge. Wisdom is organized life.
Immanuel Kant

For believe me: the secret for harvesting from existence the greatest fruitfulness and greatest enjoyment is - to live dangerously.
Friedrich Nietzsche, *The Gay Science*, Section 283

All you need in this life is ignorance and confidence; then success is sure.
Mark Twain, *Letter to Mrs Foote*, Dec. 2, 1887

Planning and developing and organising MaT projects in formal design programmes require considerable attention to detail. Quite apart from the issues canvassed in Section 2.5, there are matters relating to where and when MaT project testing is to be administered. In addition, design teams must have sufficient opportunities for prototype field-testing. Competitions involving head-to-head contests need appropriate space with their constraints and limitations taken into consideration in the specifications. Where MaT devices are to be tested out of doors, weather conditions can seriously influence performance. Probably the most important element of MaT project planning is to permit sufficient creative freedom in the specification to capture the imagination of participants and to promote engagement with the problem. As well, MaT projects tend to require with substantial early-mentoring. Consequently, design teams should be given ample opportunity for interacting with designated mentors.

E 3.1 *Bungee - omelette*: Because most readers of this book are likely to have some design experience, it is instructive to consider how a MaT project might be designed in the context of a *Product Planning Matrix* (see Sections 2.2 and 2.3). Based on a QFD style product planning matrix, develop a specification for a MaT project. The project is to use a selection of elastic bands to act as bungee cords in delivering a fresh egg from (say) a height of 10 m to the ground safely. The criterion of performance is the minimum distance from the ground reached during successful delivery of the egg (i.e. the closer to the ground the better).

3.1 MaT Projects Using Static Structures

Typical examples of a MaT project involving static structures include load carrying structures with deflection, buckling or rupture under load as the dominant mode of failure. The criterion of performance is usually associated with the magnitude of the maximum load at structural failure, the maximum structural deformation or the ultimate load to weight ratio of the structure at failure.

Design for Axial Loading (Struts)

Terminology

Relatively *slender* structures, where compressive axial loading dominates all other forms of loading, are commonly referred to as columns or specifically *struts*. In this way we can differentiate between axial loading in tension (*ties*). The common form of failure for struts is *buckling*, a type of elastic instability.

Key dimensions of struts are length L, cross-section area A and second moment of area I_{zz} about an axis normal to the long axis of the strut. A more substantial description of column behaviour is provided in Appendix A1 but, for MaT projects, buckling is the failure mode to be avoided.

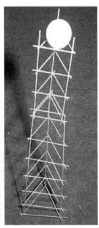

Figure 3.1 Newsprint column

Figure 3.2 Spaghetti column

Figure 3.3 Dynamic stability test for structure

Axially loaded structures may be of single-element or multi-element construction. Figure 3.1 shows a single element structure composed of newsprint. Figure 3.2 is an example of a multi-element structure using spaghetti. Figure 3.3 shows a type of simple stability test that may be specified for such structures. With single-element structures design performance is often a function of overall load bearing capacity and local crushing strength of the component. With multi-element structures performance often depends on joint behaviour. Hence, as a general rule with multi-element structures careful consideration should be given to joint design.

Performance Criteria

Design challenge results from trying to achieve best performance relative to some specified design criteria. The following performance variables may be used to describe the performance of these types of structures:

- Overall height H, measured from some base supporting structure such as a table;

- The load being carried, commonly referred to as the *payload* W_p;

- Total mass of the structure, M, excluding the payload;

- Performance criterion might be specified in the form of $(H \times W_p / M)$.

Additional performance requirements might be specified for stability of the structure while loaded. A useful type of stability criterion is the side-impact test. Figure 3.3 illustrates the nature of this test. A dynamic side-loading test is shown. The plastic or wooden bar of specified length L (typically a standard ruler) is suspended between thumb and forefinger and swung through a specified angle of arc α. In this way the stability of the structure may be tested. Alternative testing procedures may be used, as long as in each case the ground rules for testing and acceptable performance limits are clearly specified.

Design for Transverse Loading (Beams)

Terminology

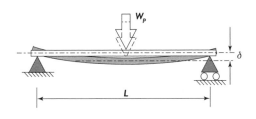

Figure 3.4 is a schematic view of a beam structure designed to carry a transverse load. A beam has span L between supports, a payload W_p and a deflection under loading of δ as shown. Two regular forms of this type of structure are used in projects. The first form specifies a given load W_p and asks for

Figure 3.4 Schematic view of a beam structure

maximum span L, with some limit on the deflection under load. The second form specifies a span L and asks for a maximum load W_p, also with deflection limits specified.

The nature of the supports imposes further constraints on beam type structures. In Figure 3.4 the beam is *simply supported*, implying that the beam canot resist any bending moments at the supports, and that the supports (at least ideally) can't impose any frictional or shearing loads on the beam. The corollary to this specification is that the supports are unable to carry any other than vertical loads parallel to (and opposing) the direction of W_p.

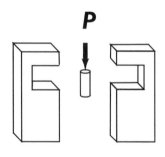

Figure 3.5 (a) Loading conditions specified for a beam structure with the load and supports in the same plane, but not at the same level

Figure 3.5 (b) A sample beam at rupture load in a testing machine. Construction materials specified were urethane foam and fiberglass reinforced epoxy resin coating

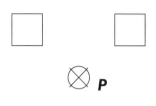

Figure 3.6 (a) Loading condition specified, with load and supports noncoplanar

Figure 3.6(b) A sample beam at rupture load in the testing machine. Construction material specified was balsa wood

Figure 3.7 An alternative solution example to the loading conditions specified in Figure 3.6(a)

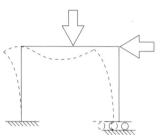

Figure 3.8(a) Schematic sketch of a frame structure and its complex deformation shape

Figure 3.8(b) An example of a balsa wood frame structure with an implied exclusion zone

Although this is the most common form of support specified for beam type MaT projects, there are alternative types of specially designed supports. Typical examples are shown in Figures 3.5 and 3.6, where the support may be used to carry moment loads. A further variant of beam type projects might specify that the payload and supports are noncoplanar.

The structures shown in Figures 3.6(b) and 3.7 are both solutions to the same set of loading and support constraints. Clearly, even with such a simple specification, thoughtful designers may devise surprisingly different solution alternatives. In the specific exercise, two supports were provided, but participants were not restricted to making use of both supports. The elegant solution structure of Figure 3.7 is not only simpler than the one in Figure 3.6(b), it is also lighter, and hence a more efficient performer, when the specific performance criterion for the project is payload-to-mass ratio.

Yet another type of structure, suitably engaging for MaT projects, is that shown schematically in Figure 3.8(a). Due to the complex behaviour of frames, these types of structures are usually met and analysed by engineering students in later parts of the course. When faced with practical frame design, in the absence of suitable analytical understanding, design teams are virtually forced to perform substantial field testing with the materials specified. In addition, there may be specified "exclusion zones", into which the structure may not encroach. Figure 3.8(b) is a sample submission for a frame type project.

Performance Criteria

The performance of beam structures is determined by the span L between supports, payload W_p and total mass of the structure M excluding the payload. Often in MaT projects the material of construction is pre-determined by the design specifications. Consequently, the design problem will focus on either a given payload W_p to be carried over a *maximum span*, or a *given span* beam carrying a maximum payload $(W_p)_{max}$. Ultimate performance might be specified in one of two possible ways:

(a) For load limited structures, a possible performance criterion is $(L \times W_p /M)$. In this case the structure may be restricted to a limiting deflection δ. This approach is a reflection of structural specifications in buildings, where beam deflections are usually limited to some small proportion of the span. Beams used in buildings are designed to carry the applied loads without clearly visible deflection. A generally aceptable architectural limit on beam deflection is about 0.35% of the span, as this is not easily detected by visual inspection. There may be other restrictions that could limit the allowable deflection in a MaT beam project, but the above architectural limit provides an order of magnitude starting point for load-limited beam specifications.

(b) For deflection limited structures a possible performance criterion is $(L \times W_p)/(M \times \delta)$.

3.2 MaT Projects Designed to Use Moving and Dynamic Devices

Typical examples of MaT projects using moving or dynamic devices invariably involve energy conversion and the appropriate matching of loads and energy supplies. All of these projects are concerned with energy conversion and the key elements of design are input energy sources (thermal, potential, kinetic, elastic, electrical, solar, nuclear) and desired output performance variables (speed, mass deployment, ballistic function) coupled with specific combinations of constraints and restrictions.

Energy Sources

- *Direct heat*: includes flame from burning; candle, gas blowtorch, match, most flammable materials.

- *Indirect heat*: includes electric element, incandescent light globe, hair dryer, solar heating, heated fluid or solid, steam.

Potential Energy (Gravity Driven)

- Virtually any *mass raised to some height*: tennis ball, golf ball, coins of various denominations and quantities, marbles, beverage-filled containers (cans and bottles), dry food stuff (beans, peas, wheat), lead shot, brick, sand, metal weights - gravity is identified as the driving force, because in a general sense any form of stored energy is regarded as potential energy.

Kinetic Energy

- These sources are *usually converted from some other form of energy*, but the kinetic manifestation may be easier to measure and specify than the source energy: air or water jets, pendulum impact, power tool output (power drill, power stapler).

Elastic Energy

- Elastic energy is obtained from *strains imposed in flexible materials* and *fluid under pressure*: elastic bands, balloons, mousetrap springs or springs of any kind, compressed air in beverage containers, gas or fluid contained in spray cans, bicycle pumps.

Electrical Energy

- *Direct electrical energy* may be used from the local electric grid. In general this type of energy is excluded from MaT projects; it may be used to drive ancillary equipment, such as power tools, air compressors, or hair dryers (for thermal energy). More commonly, however, batteries are used for the supply of energy to drive electric motors. Hobbyist motors may be used as a means of converting electrical energy to mechanical energy or the reverse of driving the motor mechanically (by elastic energy, for example) to obtain electrical energy. This latter approach is very inefficient and is substantially inferior to dry-cell batteries (which are a form of chemical energy).

Solar Energy

– *From sun or wind driven systems*: There are some serious limitations to the use of solar energy in MaT projects. The solar cells are expensive and the project entries must be tested in suitable sunlight. An alternative form of solar energy is through the use of wind generators for electric energy. This too can be costly and awkward for a relatively brief project.

These limitations should not necessarily discourage the use of solar energy, either as a cell, or as a source of heat energy in such projects. Nevertheless, the application of solar energy will require considerable ingenuity and planning if it is to be used in a MaT project.

Nuclear Energy and Chemical Energy (Other than Batteries)

– As previously noted, one overriding concern in specifying dynamic make-and-break projects is the operational safety for all participants. This alone precludes the use of any radioactive or indeed any form of hazardous materials. Occasionally, enterprising students will resort to the use of explosive energy from gunpowder (once readily available from fireworks, but now mostly restricted) or homegrown mixtures of explosive chemicals. This should be aggressively discouraged when specifying the restrictions associated with projects.

Potential Performance Requirements and Variables

Wheeled Vehicles

– These types of vehicles may be used to deliver a payload over some specified distance or to perform distance or load endurance tests. The design challenge lies in finding the most effective means of converting the available energy into the desired output performance. The design usually tries to match the available input energy to both the conversion technology used in the vehicle drive system and the overall energy requirement of the desired output.

Figure 3.9 One solution to a balloon powered vehicle MaT project in 1982

Figure 3.10 A mousetrap driven vehicle design

Figure 3.11 An alternative balloon vehicle design

Figure 3.12 The mousetrap vehicle winner of the National Engineering competition in 1985

As an example, consider the potential energy available from a golf ball delivered from some height above the operating surface. This energy is to be used to drive a wheeled vehicle with a payload of one golf ball. Further suppose that for this vehicle the farthest distance covered in a straight line over level ground determines the best performance. With this type of project the key design challenge becomes that of capturing the potential energy of the golf ball and converting it to the kinetic energy of the vehicle. Because speed is immaterial in this case, slow release of the captured energy will result in smaller losses and a more efficient energy capture or release system. In this way we can avoid the inherent losses resulting from applying direct impact energy to drive the vehicle. This very brief initial conceptual analysis of the potential to kinetic energy conversion problem is sometimes referred to as the *matching process* in design.[3.1] Figures 3.9 through 3.12 show some examples of wheeled vehicles designed and constructed by MaT project teams.

Performance Criteria

– Wheeled vehicle performance is defined by payload W_p, total mass M of the vehicle, excluding the payload, the distance S traversed and the time T taken to perform the task. These measures are all output, or performance, variables. The best performance may be defined as the vehicle with the largest value of a performance index of $S \times [1 + (W_p/M)]/T$. For a surprisingly large range of variants of this project, this performance index may be specified with any combination of the performance variables held constant.

This performance index has the clear advantage of being able to be evaluated "on the fly", as the performances of competing designs are tested, providing an engaging head-to-head challenge for participants.

Vehicles Designed to Travel in or over Water

– Vehicles designed for water have strong similarities to wheeled vehicles, with the added complication of matching energy input to propulsive technology and navigation in water. In addition, for underwater vehicles there is the further complication of sealing out water during travel. Almost all learning institutions have some form of water feature that lends itself to this type of design project.

As an example, consider the available energy source provided by a dry cell battery, driving a (specific) miniature electric motor, available from most hobbyist suppliers. To maintain a level playing field for participants, the overall size and material of the vehicle "hull" may be specified in the problem statement, or provided as part of the project "givens". Distance tra-

---

3.1 See also Chen and Hoffmann (1995); Narri and Mummadi (1999); Du and Nair (1996); Allison and Cole (1993).

Figure 3.13 Sample submission to an underwater vehicle MaT project

Figure 3.14 Underwater MaT project vehicle in action

Figure 3.15 Hovercraft entry for airborne vehicle MaT project

Figure 3.16 Winged vehicle entry for airborne vehicle MaT project

versed in a straight line and the maximum average velocity could determine the best performance. Figures 3.13 and 3.14 are samples of design submissions for a MaT project asking for the design and construction of an underwater delivery vehicle.

Performance Criteria

Again, using the output performance variables defined for wheeled vehicles, the best performance may be defined as the vehicle with the largest value of a performance index of $S \times [1 + (W_p / M)]/T$. *Matching* is the early conceptual evaluation of the best means for transforming the available energy into propulsive energy. With water-based designs this matching process requires either an understanding of water propulsion technology or some early experimentation. Additional performance criteria might be the mass of water incursion during transport and various navigational issues, including vehicle guidance.

Airborne Vehicles

– Airborne vehicle performance is guided by matching the energy source to airborne propulsion technology. Propulsion in these types of vehicles involves both forward thrust and lift to maintain the vehicle aloft during transport. Two principal types of designs are available in this classification. The first type, by far the easier of the two, deals with ballistic propulsion technology. The second type involves designs that remain aloft for some time trial.

Performance Criteria

Using the previously defined performance variables, best performance is identified by the largest value of $S \times [1 + (W_p/M)]$. With ballistic vehicles the time of flight is usually very short and difficult to measure and, in any case, it is unlikely to be of importance in the assessment of performance. In a time trial of airborne endurance (*i.e.*, how long the aircraft remains aloft) the performance index becomes $[1 + (W_p/M)]/T$. Figures 3.15 and 3.16 show design submissions to a MaT project based on the design and construction of airborne endurance vehicles.

Devices with Mixed Performance Requirements

Design opportunities in this category are almost limitless and depend only on the ingenuity of MaT project planners. Some examples follow.

(a) Conversion of energy, for example, using direct heat to raise a mass

Consider using the heat available from a candle to raise a mass. The challenge is to design some appropriate (most effective) means of capturing the available energy and converting it to mechanical energy in some device to do mechanical work. The best device will raise a mass M to a height H, in minimum time T, with a performance index of maximum $M \times H/T$. There are many different forms of this type of project, using the whole range of possible energy sources and operating conditions.

Matching in these exercises is concerned with the most efficient means of converting the available energy into mechanical work. In the case of the candle-burning problem some form of material expansion used as leverage, or steam-raising are excellent performers. It is instructive to do some early conceptual evaluation of the many other possibilities. This approach will help weed out candidate designs that could impose undue constructional difficulties, unnecessary heat loss or mechanical penalties.

(b) Using the vehicle's structural properties to prevent damage to the payload

Probably the most often used example is the safe transport of a fresh egg from some height to ground level, using a vehicle made of newsprint. This exercise is well known internationally as "Project Egg-drop". A wide range

Figure 3.17 Three entries to the "egg drop" MaT project, a parachute, an elastically constrained space-frame and an elegant energy absorbing cone of paper - all were successful in safely delivering the egg from a height of 9 meters to ground

Figure 3.18 Two submissions to a MaT project requiring the conversion of potential to kinetic energy. The two designs on the right use the impact energy of the falling golf ball, a very inefficient energy conversion proces. The design on the left uses the energy of the falling ball to extend an elastic which is subsequently used to drive the vehicle. This approach is much more efficient than direct conversion of impact energy to kinetic energy

of materials, other than newsprint, may be specified and, in some forms of this exercise, a delivery target location may be also specified. The newsprint material may be used as parachutes, airborne delivery vehicles or, as is most common, strain energy absorber by crushing of the paper vehicle components. Clearly, in this case the performance index is partly a binary measure of failure (egg breaks) or success (egg survives). Further performance criteria may be imposed such as payload to mass ratio, or constructional elegance.[3.2]

3.2 See, for example, Jones and Wang (2001); www.angelfire.com/md/mccscience/eggdrop.html; www.college.hmco.com/education/pbl/project/project3.html#problem;
home.neo.rr.com/physicsisphun/egg_drop.html ; www.geocities.com/r_dman2000/egg_drop_project.html;
www.glenbrook.k12.il.us/gbssci/phys/projects/q2/ecrub.html;
www.silverfalls.k12.or.us/foxes/staff/read_shari/egg_drop_project.html.

(c) Head-to-head comparison of energy conversion rates

Drawbar pull is the term used for the force generated at the tow-bar location of a vehicle. This pull, multiplied into the distance travelled by the object on which the pull is exerted, defines the work done by the vehicle. A useful way of demonstrating drawbar pull is to attach two competing vehicles by their respective tow bars and make each vehicle attempt to drag the other over some specified distance, much like tug-o-war teams do when pulling on a rope.

Drag racing, probably familiar to most motoring enthusiasts, is a head-to-head race against time of two vehicles over some specified distance. Previously defined performance indices may be used, but the head-to-head race is often more engaging for the participants.

Figures 3.17 and 3.18 show sample submissions to an *egg-drop* project and to a project requiring the conversion of potential energy of a falling golf ball to the kinetic energy of a vehicle.

3.3 Some Key Materials of Construction

(a) Static MaT Projects

With static, load-carrying projects the nature of the material should not require access to specialist tools of construction. Typically, prescribed materials should be readily available from toy or hobbyist stores, art suppliers, supermarkets or the local hardware store. Balsa wood, foamcore, spaghetti, string or rope, elastic bands, balloons, drinking straws, corrugated cardboard and bamboo skewers fall into this category. Perhaps the simplest material of construction for static projects is the humble newsprint. Occasionally polymers such as styrene or urethane may be explored in static projects. These materials are only available from specialist suppliers and, in general, procurement of material for projects using these materials involves considerable staff resources. All of these materials may be worked with readily available household implements such as scissors, razor blades or utility knives.

Most of these materials may be joined with hobbyist glues, with appropriate adhesive tape or cotton thread, although some might require special preparation to achieve a strong bond. The essence of material choice is to make it easily accessible and easy to use as a construction material, with readily available hand tools, by participants who may have varied skills and background experiences.

In most conventional engineering design problems the properties of materials of construction are either specified or readily available from material suppliers. What makes static MaT projects interesting and fresh is the use of unconventional engineering materials. For most of these unconventional materials, such as balsa wood, newsprint, soda-straws and spaghetti, mechanical properties are not readily available. Variations in structural performance of these unusual structural materials can substantially influence performance of the load-bearing structure. Consequently, students should be encouraged to experiment with their construction materials to establish performance characteristics that might lead to a best (most optimal) design. These experiments should form part of the total MaT project and should only use apparatus readily constructed from simple materials around the home. Examples of simple test equipment and test procedures are described in Chapter 4.

(b) Dynamic MaT Projects

Materials used for constructing dynamic MaT projects are virtually unlimited. As a general rule, with these projects creative freedom should be free from restrictive material choices. Where the project needs to be constrained to level the playing field for participants, some specific materials of construction may need to be supplied. Typical examples are the *Payloads-on-Pools* project (*PoP* – 1975/1983) and the *Beverage Can Roller* (*ABC* – 1991).[3.3] With the *PoP* project, where a water-based delivery vehicle was to be designed, the boat hull and the drive engine were provided to student teams. With the *ABC* project, the design called for a drive system that had to fit into a beverage can and, in that case, the beverage cans were provided for student teams.

These projects offer a large variety of opportunities for energy conversion. The key criterion of choice for this type of MaT project should be the safety of the participants. Unrestrained explosions, explosive conditions or any other forms of hazardous conditions should be avoided. Safe conditions of operation should be the main guiding principle in the planning of these types of projects. Unfortunately, in some MaT projects (static or dynamic), hazards may arise due to circumstances unforeseen at the planning stage. Some examples of unforeseen hazards are described in *Daring Young Designers and Flying Machines* (*DYDaF* – 1973) and in *Air Lift Engine* (*ALE* – 2000).

In addition, the energy sources specified should be readily available to participants and, just as with static MaT projects, the project specifications should avoid the need for specialist manufacturing skills or equipment.

3.4 Level of Difficulty and Complexity of Projects

Project difficulty and complexity are serious concerns for academic staff involved in planning MaT projects. With conventional engineering design projects past experience is often a useful guide to planning the scale and difficulty of the work. In MaT projects it is relatively easy to make the problem to be solved or the work required so difficult that the students involved might simply choose to avoid the problem and submit trivial or "bypass"[3.4] solutions. Clearly, this result is counterproductive to the basic objective of

---

3.3 For a full description of these projects refer to Chapter 7.
3.4 See Example I - airborne vehicle below.

MaT projects, namely early engagement of student interest. A useful guide for planning engaging MaT projects involves solving the problem as a *gedanken*, or thought experiment, by the project planner. At least three alternative solutions to the specified problem should be considered, with solution deliveries well within the skill capacity and time constraint of the participants. In this way appropriate guidance may be provided for students unable or unwilling to get started with such a project.

One of the inescapable difficulties of MaT projects is that of overcoming the fear of public exposure of failure. Some students will identify with this aspect of the project to a level where they may be unwilling to commit to a course of action. Even with well-planned MaT projects, some students may simply trivialise and avoid the problem, because *"We are unlikely to be asked to something as silly as this in real life."* Care in planning and continuous guidance, available throughout the project, are essential elements of the whole educational experience. The case studies of Chapters 5 and 7 provide examples of MaT project planning and management.

The engaging freshness and immediacy of MaT projects are considerably enhanced by both design teams and academic mentors, experiencing solution discoveries (the *aha!* experience) as the project unfolds. This process makes the project a valuable (and valued) learning experience for both sets of participants. Naturally, there are unforeseen project difficulties and complexities that accompany such a learning process. Some instructive samples of project difficulty and complexity are indicated in the following examples:[3.5]

1. Airborne transport vehicle required to remain aloft for at least 10 seconds (*DYDaF* - 1973). Performance index was $W_T \times T/W_D$, where W_T was the total vehicle load including payload, W_D was the tare weight of the device, and T the time that the device would stay airborne. Materials of construction were unlimited, but proprietary models, purchased from model shops, were excluded.

   • The major difficulty of this project originated from the requirement of being airborne for a relatively long period. The common solution attempted by participants was a glider type of vehicle. In international competitions of superlight gliding vehicles the required flying time of 10 seconds is achieved easily.[3.6] With relatively inexpert construction techniques this difficulty was almost insurmountable with a glider. In addition, there were substantial limits on the range of gliding distances available in an indoor test environment. Weather conditions that can substantially influence performance precluded outdoor testing. Consequently, the best solution to this problem, when set to be tested in a limited space

---

3.5 Refer to Chapters 5 and 7 for documented case studies of some of these examples.
3.6 The current international record for superlight gliders is 64 seconds held by Danjo and Ishi; refer also to Section 7.3.

indoors, was a ground-effect device, such as a hovercraft (see Figure 3.15). Once this solution opportunity is recognized, then constructional difficulties become the next major hurdle to be surmounted. Only few engineering students have the necessary skills to design and construct such a device. A "bypass" solution offered to this problem was a piece of dry ice. This "solution" failed because the gas film on which the dry-ice "vehicle" floated was so minuscule that imperfections in the floor interfered with transport performance. One student team avoided the problem completely by submitting a bird feather with a single lead shot attached as payload.

This example is cited as a case of poor, and personally biased, planning. Clearly, the initial *gedanken* evaluation of possible design outcomes did not reveal the inherent difficulties that students might experience in this project. A degree of complexity was added by the aerodynamics aspects of the problem. Most relatively new engineering undergraduates would not have had the experience or learning necessary to manage the evaluation of the aerodynamics involved in this MaT project. Personal bias guided this project because the academic suggesting the problem had extensive expertise in building model aeroplanes.

2. Submarine transport vehicle to carry a payload over some distance while submersed (*UWACS*–1988). In this project the performance index was $W_T/(T \times W_D)$ with the symbols having the same meaning as those for Example 1 above, but T here is the time taken to cover the specified test track distance.

   • The major sources of difficulty arose from the need to seal the vehicle against water incursion during transport, as well as the proper adjustment of buoyancy. Some of the offered solutions simply rolled along the floor of the swimming pool where the devices were tested. None of these nonfloating vehicles performed as well as the floating submarines (Figures 3.11 and 3.12).

3. Vehicle to be driven by the energy available from three or less toy balloons.(*Huff-n-Puff* - 1982) Nominated performance criterion was $W_T/(T \times W_D)$. Testing was to be performed out of doors.

   • Design teams in this project soon realized that wind conditions would seriously influence performance of the vehicles using exposed balloons. In fact, the larger the balloon (to get as much air and associated energy density into the balloon as possible), the more serious became the influence of wind resistance on performance. Some participants chose to reformulate the problem specifications and used the balloon simply as a diaphragm between air and water in a closed vessel (*e.g.*, a 1 litre plastic beverage container). This "solution" essentially contravened and, in fact, bypassed the original intentions of the project. Experience with this project highlights the need for maintaining creative freedom in dealing with (possibly unforeseen) MaT project "solutions". In order to maintain student enthusiasm generated by MaT projects, the entries were classified in several categories. Although this experience is most valuable for MaT project planners as well as participants, it also points up the inherent weakness of specifying designs without a carefully planned follow-up consultation process. The results can also provide valuable aha! experiences with the unforeseen, or, in some cases, surprise solutions.

- Difficulties arose with this project due to the unpredictable environmental influences on performance as well as the issue of poorly constrained specifications. Specifications should not become so over-constrained that they stultify creative freedom, however, as a measure of competent planning, all participants in the exercise should be made aware of the clear spirit of the exercise. Ultimately this can only occur through carefully planned and administered consultation between designers and academic "clients".

4. Design teams were given three pieces of 3 mm thick balsa wood, each measuring 900 mm x 150 mm, and were asked to construct a "bridge" to carry only its own weight *simply supported*. No other materials of construction were permitted. Criterion of performance was the length of the "bridge" (*i.e.*, the longer the better).

- A common construction type, used by designers tackling this problem, involved simple triangulated structures with balsa wood "joints". Only two design teams chose to use a "suspension bridge" structure, slicing the balsa into thin slivers and joining the slivers with tiny blocks of balsa. One team used a jointing block of balsa with a hole through the centre, thereby providing a friction joint. The other team used a modified version of this friction joint, but, in essence, both teams had similar design solutions to this problem. Figure 3.19 shows the two types of friction joint used for the "string suspension bridge" solutions. The length with

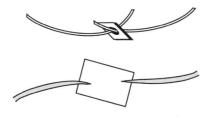

Figure 3.19(a) Schematic sketch of one type of friction joints used in the balsa "string suspension bridge"

Figure 3.19(b) Schematic sketch of the other type of friction joints used, where the slivers of balsa were wedged into a small block

these suspension solutions was only limited by the length of the building in which the device was constructed. (Note that this is a slightly extended version of the "paper bridge" project described in Chapter 2, *P2.1*).

- A major source of difficulty arose with this project due to unwillingness of participants to substantially modify the materials of construction provided. (Refer to the *lateral thinking* examples at the end of this chapter.)

3.5 Team Dynamics

MaT projects are tackled by groups of designers working together as a team. The implication of "teamlike" behaviour suggests group members working together coherently towards some shared goal. Because this experience is likely to be the first time that student designers work together in a team larger than two members, some team dynamics issues need to be addressed in the planning of the project. The social dynamics of teams involved in project-based learning has been studied by educational psychologists, but not commonly encountered in research by engineering educators.[3.7] This presents a mildly concerning situation, when we recognise the number of engineering projects offered to groups of students in tertiary institutions. Teams may be chosen by "natural selection" of members, who may have had some history of cooperative work in previous projects. In some instances an arbitrarily chosen group of designers may be asked to work as a team. Team members can undertake many roles. They can act as :

- *Ideas person* – responsible for offering and/or eliciting ideas from the team;

- *Detail follow-up* – person responsible for overseeing the early development and conceptual evaluation of ideas into realistic design plans;

- *Information resource* – someone responsible for collecting and storing information needed to develop the design solution;

- *Timekeeper* – someone responsible to ensure that deadlines are set and met during project execution;

- *Secretary* – a note taker to maintain records of meetings and discussions.

Team leadership may be exercised by some, or all, of the persons taking these roles. In the song about General Jubilation T Cornpone, an invention of the "Li'l Abner" comic strip artist Al Capp, leadership is satirised by the words of Gene De Paul in the musical:[3.8]

3.7 Moje et al. (2001); Werner and Lester (2001); Connor-Greene (2002).
3.8 *Li'l Abner: The Musical*, Gene De Paul and Johnny Mercer, 1956; Al Capp's *Lil Abner: The Frazetta Years*, 1960–1961, Al Capp and Dennis Kitchen.

When we fought the Yankees and annihilation was near,
Who was there to lead the charge that took us safe to the rear?
Why it was Jubilation T. Cornpone;
Old "Toot your own horn – pone."
Jubilation T. Cornpone, a man who knew no fear!

When we almost had 'em but the issue still was in doubt,
Who suggested the retreat that turned it into a rout?
Why it was Jubilation T. Cornpone;
Old "Tattered and torn – pone."

Jubilation T. Cornpone, he kept us hidin' out!

Project team leadership is exemplified by anyone in a team prepared to "show the way" in progressing the project and its ultimate solution. A good team leader will ensure that all members of the team are given the opportunity to respond to the needs of the team. However, care is needed in planning the project to elicit fair sharing of work and creative input. In most cases a team leader may be appointed to oversee the operation of the team. Some students rise to the opportunity of leadership, but in many cases careful mentoring is required to permit the development of the shared goals of the team. Occasionally students will prefer to work in a "democratic structure", with no clear leader. However, academic staff, acting as project mentors, can help the team to recognise early in the project that project delivery in engineering is rarely a democratic process. An overwhelmingly important feature of well-planned MaT projects is the shared acceptance that any design conflict that needs to be cleared away before construction starts should be done so by consultation with design mentors.

3.6 Morphology of MaT Projects[3.9]

Table 3.1 is a simple morphological chart of static load-carrying type MaT projects and Table 3.2 is a morphological chart for dynamic MaT projects. These tables can provide a useful starting point for generating (possibly new) MaT project ideas. Since opportunities for new project ideas are guided only by the ingenuity and inventiveness of the project designer, interested readers may add many more items to the tables.

> **E3.2** *Energy conversion* – In the description of the golf ball driven vehicle (see Figure 3.16) it was noted that using the impact of a falling golf ball to drive the vehicle is less efficient than using the potential energy converted to strain energy of an elastic to drive the vehicle. Carry out a simple analysis to prove or disprove this statement. List all the types of losses incurred in both approaches to the conversion of energy from that of the falling golf ball to the energy of the moving vehicle.

3.9 Morphology is the study of the form of objects and abstract systems. In design it has become synonymous with a procedure that permit the generation of new designs through the combination and re-combination of different design attributes.

Table 3.1 Static structures used for carrying loads					
Nature of structure	Complexity	Geometry and nature of load application	Materials of construction	Adhesives	Performance criterion
Beam	Simple, possibly single element	Planar loading	Newsprint	None	Maximum load carried
Column	Multi-element, triangulated	Planar loading	Foamcore	Adhesive tape	Maximum payload
Torsion-bar	Complex non-linear shape elements	3-D loading	Balsa wood	Balsa glue	Lightest structure for a given load
Combination of two or three types of structures[1]	Could be very complex	Impact load-ing	Fibreglass-resin composite	Resorcinol	Survival of protected payload

Note 1: Some structures involve the combination of more than one type of loading. An example is a "twisting beam" where a component loaded in bending has a twisting moment imposed on it (see Figure 3.6).

E3.3 *Thinking outside the square* – Some special types of projects are particularly well qualified for the description of thinking outside the square, or *lateral thinking*,[3.10] in which some casting off of perceived constraints can lead to surprisingly elegant outcomes. One such project is the *A4-Paper Bridge* project *(P2.2* in Chapter 2*)* and two others are, the *Balsa Bridge* project (Figure 3.19) and the *Twisting Beam* (Figure 3.6). The two bridge projects can yield surprise solutions if we modify (ever so slightly) the notion of how the material of construction may be used, or the way the supporting structure may be modified. The solution to the twisting beam yielded the surprisingly elegant solution shown in Figure 3.7 due to a re-interpretation of the way the supports were to carry the loading.

The two well known problems, often used to forcefully demonstrate the process of thinking outside the square, are the *Nine Dots* problem and the *Six Matches* problem. Figures 3.20 and 3.21 show these *mind-benders*. The solutions to these problems are interesting examples of how design specifications are sometimes interpreted as perceived physical constraints. There are other problems that exhibit this type of behaviour and a further extension of this exercise is that one should try to uncover and categorise as many of them as possible. Some samples are listed in the appendix.

3.10 Edward de Bono (1970).

Table 3.2 Dynamic MaT projects					
Type of device	Energy source	Test conditions	Materials of construction	Payload	Performance criterion
Delivery vehicle	Candle	Indoor - smooth surface	Any	Any	Speed, payload
Lifting mechanism	Incandescent light globe	Outdoor - brick paving	Balsa wood	Fresh egg	Safe survival of payload
Flying device	Mass from given height	Outdoor bitumen	Newsprint	Delicate instrument	Safe delivery of payload
Sorting mechanism	Elastic band	Shallow pool	Polymer	Human	Speed, water exclusion
Clock mechanism	Bungee cord	Deep pool surface			Distance travelled
Demonstration of physical concept[1]	Helical torsion spring	Vertical rope			Ingenious construction
	Helical tension spring	Vertical smooth wall			Constructional elegance
	Mousetrap	Horizontal wire			Time of flight in air
	Size D battery	Airspace over grassy plain			Rate of energy conversion
	Compressed air from hand pump into beverage bottle	Deep pool under water			Time taken to perform task
	Solar energy				
	Balloon				

Note 1: Demonstration of a physical concept is an "open-ended" design project[3.11] and, as the name implies, it does not limit the nature or scale of the project. The main benefit of this approach is one of offering creative freedom largely untrammeled by the constraints of design specifications. An unfortunate weakness of such an open-ended project is possible student perception that any submission might suffice as long as it meets the definition of "physical concept". Using this definition one might find a submission of a simple string and mass pendulum. Additional instructions in the statement of such a MaT project might ask for the "demonstration" to be dynamic and last (say) for one minute and to employ (say) a standard golf ball to trigger the demonstration. These added instructions may increase the complexity of a project, but it still retains its open-ended nature.

3.11 See, for example, Samuel and Lewis (1974); Samuel (1986); demonstration of a physical concept is MaT project Metaphor—1984, more fully described as a documented case in Chapter 8.

(Note 1 continued)

Although the general discussion of open-ended design projects is outside the scope of this book, a further difficulty arises with project Metaphor. This difficulty is associated with the great variety and level of understanding of physical concepts. Because this type of MaT project offers engaging design opportunities, it is useful to further constrain the problem by specifying some given physical concept, such as (say) resonance or Coriolis' acceleration.

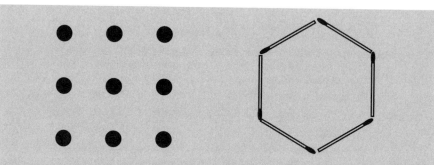

Figure 3.20 Nine dots: Draw four contiguous line segments that pass through the centre of all the dots without lifting the pencil from the paper or retracing a path

Figure 3.21 Sic matches: Form four equilateral triangles using only six matches without breaking any of them

E3.4 Tables 3.1 and 3.2, respectively, show attribute lists for static and dynamic MaT projects. New combinations of attributes may be assigned to as yet unexplored MaT projects by connecting various cells in the table across the page. For example: one "new" MaT project might involve (refer to Table 3.2) a "sorting mechanism" for two different sizes of Easter eggs, simulated by tennis balls and ping-pong balls all mixed together in a bin. The device should be capable of sorting two tennis balls and three ping pong balls into buckets, using energy available from either rubber bands or helical springs. Any material of construction may be used. Prepare an initial evaluation (with sketches of ideas) of possible design solutions to this design problem.

E3.5 You are shown ten bags, filled with marbles. You are told that the number of marbles in each bag differs, but all bags contain at least fifteen marbles. Nine of the ten bags only contain marbles weighing 10 grams each. One (faulty bag) of marbles contains marbles weighing only 9 grams each. Design a weighing scheme that permits the discovery of the faulty bag in one single weighing.

E3.6 A large rectangular painting is to be supported by two picture hooks above it. The painting is supported by a length of wire, with its ends securely attached to the two upper corners of the painting. Design a means of hanging the painting so that if either one of the two picture hooks is removed from the wall the painting will fall down. The hanging wire may be made as long as is necessary, but once the painting becomes unattached from either hook it must, as a consequence, also become fully unattached from the other hook.

E3.7 Your design team is provided with a number of wooden blocks, all of equal size and weight. Your task is to build a "tower" using these blocks such that the inner edge of the uppermost block is entirely outside the edge of the table on which the blocks are being balanced (see Figure E3.7). Your objective is to achieve this state with minimum numbers of blocks. Can you estimate this minimum number n?[3.12]

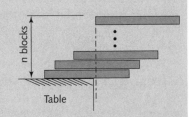

Figure E3.7 Blocks balanced on the edge of a table

E3.8 A right triangle, composed of four geometric parts is shown on the top arangement in Figure E3.8. When these same geometric parts are rearranged as indicated in the lower figure, a unit square is left empty. Where does this extra space come from?

E3.9 *Tangrams*[3.13] – Reputedly an ancient Chinese invention, tangrams have been used by teachers to enhance geometric visualisation in children. Based on a very simple set of seven geometric pieces, the tangram puzzles that may be constructed from the basic tangram pieces is quite remarkable. Figure E3.9 shows the seven basic tangram pieces together with relative dimansions. As a simple test of geometric visualisation construct a 4 x 4 square using all the seven pieces.

The tangram is a two-dimensional spatial visualisation exercise. A three-dimensional version is the *Soma Cube* puzzle.[3.14] This is a cube of edges three units long, with the twentyseven unit cubes composed into seven unique components. Carry out some basic research and discover why there are 141 possible unique constructions of this cube.

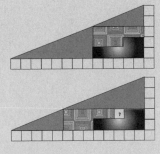

Figure E3.8 Rearranging a triangle

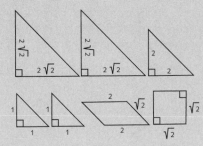

Figure E3.9 Tangram pieces

3.12 See also Johnson, (1955).
3.13 See, for example, www.ex.ac.uk/cimt/puzzles/tangrams/tangint.htm
3.14 See, for example, web.inter.nl.net/users/C.Eggermont/Puzzels/Soma/

3.7 Chapter Summary

In this chapter the following issues were addressed:

- The design and development of static and dynamic MaT projects;
- Some key materials of construction;
- Team dynamics and how to plan for teamwork;
- The morphology of MaT projects;
- Thinking outside the square.

4
PROPERTIES AND APPLICATION OF SOME UNCONVENTIONAL ENGINEERING MATERIALS

If in doubt, make it stout,
Use only materials you know about.
Anonymous

Engineers commonly use materials whose properties they do not properly under-
stand, to form them into shapes whose geometries they cannot properly
analyse, to resist forces they cannot properly assess, in such a way that the pub-
lic at large has no reason to suspect the full extent of their ignorance.
John Ure, 1998

Engineers commonly use conventional materials (metal, timber, concrete) for construction. In general, these materials require considerable manufacturing effort to change them from the "raw" state to the useful shape and size specified in a design. Consequently, it is always prudent to plan and generate carefully prepared drawings of parts and assemblies. This approach is based on old engineering caution. "It is always easier to rub it out on paper than in steel."

This carefully planned and persuasively prudent approach requires relatively long timeframes to proceed from idea through working prototype to final product. The colloquial term used for the lead-time required to progress through the various phases of design to final product is "cradle to launch" (CTL) time. *Agile manufacturing* is the title used to describe manufacturers with the capacity to respond quickly to changes in consumer demand, with shorter CTL times. The automobile industry, where model and style changes vary with fashion, with CTL times of five to seven years in the past, agile manufacturing now aims for CTL times of three years or less. The Ford Pinto disaster is in part attributed to the unusual rush to get that model into the marketplace in two years.[4.1] Xerox has exceeded seven years of CTL time for one of their most advanced computer-managed photocopying machines.[4.2]

The intended spontaneity of engaging MaT projects precludes such planning and long lead-times. Hence the need for some unconventional engineering materials, easily manufactured into the forms and structures required, for these short time-response encapsulated engineering design

---

4.1 See *Mother Jones* (1997); also http://www.motherjones.com/mother_jones/SO77/dowie.html.
4.2 Smith and Alexander (1988).

experiences. Some typical, easily worked, construction materials are balsa wood, newsprint, foamcore, spaghetti, soda straws, bamboo skewers and polyurethane foam. In most cases the structural properties of these materials are not readily available to the engineers involved in constructing MaT project devices. It is instructive and even desirable to carry out simple tests to establish the appropriate mechanical properties of materials used in these projects.

This chapter describes the nature, availability and mechanical properties of some useful unconventional engineering materials that might be specified for MaT project construction. Also described and evaluated are some easily available, unconventional energy sources that might be applied as MaT project prime movers. Both, materials and energy sources, need to be unambiguously specified so that performances of MaT project devices are compared, as far as possible, only on the creative use of these basic resources. For these reasons, wherever appropriate, the background and provenance of these materials and energy sources are also described.

4.1 Balsa Wood

Balsa wood (*ochroma lagopus*) has been the standard material for model airplane construction since it first became readily available in the United States in the late 1920s. Its outstanding strength-to-weight ratio enables hobbyists to construct durable models that fly in a realistic manner. Balsa also absorbs shock and vibration well and is easily glued, cut and shaped with simple hand tools.

Balsa trees grow naturally in the humid rain forests of Central and South America. Their range extends south from Guatemala, through Central America, to the north and west coast of South America as far as Bolivia. However, the small country of Ecuador, on the western coast of South America, is the primary source of model aircraft-grade balsa. Balsa trees

Figure 4.1 Colorful carved balsa birds and fish from Ecuador

need a warm climate with good rainfall and well-drained soil conditions. For these reasons, the best stands of balsa usually appear on the high ground between tropical rivers. Ecuador has the ideal geography and climate for growing balsa trees. The name *balsa* is Spanish, meaning raft, in reference to its excellent floatation qualities. In Ecuador it is known as *boya*, meaning buoy, again referring to its fine flotation qualities. Interestingly, also in Ecuador, where balsa is a comparatively cheap commodity, it is commonly used for carving and folk art (see Figure 4.1).

There are no entire forests of balsa trees. They grow singly or in very small, widely scattered groups in the jungle, reproducing from hundreds of long seedpods, which open up and, transported by wind, scatter thousands of new seeds over a large area of the jungle. Each seed is airborne on its own small wisp of down, similar to the way dandelion seeds spread. The seeds eventually fall to the ground and are covered by the litter of the jungle. As new balsa trees grow, the strongest of the new growth will kill off the weaker shoots, and as the trees mature, there may be only one or two balsa trees to an acre of jungle.

Balsa trees grow very rapidly, growing within six months after germination to about 25 to 30 mm in diameter and 3 to 4 m in height. At harvesting, about 6 to 10 years after germination, trees may reach heights of 20 to 30m and diameters of 300 mm to greater than one metre. If left to continue growing, the new wood, growing on the outside layers of the tree, becomes very hard and begins to decay in the centre. Unharvested balsa trees may grow to a diameter of more than 2 m, but trees permitted to grow for 10 or more years provide very little usable lumber. Due to its sparse growth in forests, as well as the difficult Ecuadorian terrain, balsa is usually harvested by relatively simple, labour-intensive methods.

Balsa's lightness is due entirely to its unusual cell structure. During growth, the large, thin walled cells of balsa are filled with water and lignin under osmotic pressure, and the cells behave as pressurised columns.[4.3] Lignin, the solid gluelike material in all timbers, binds the cells together. However, due to the large cell structure of balsa, only about 40% by volume of a piece of balsa is solid substance. Green balsa wood typically contains five times as much water by weight as actual wood substance, compared to most hardwoods, which contain very little water in relation to wood substance. In order for it to become commercially useful timber, green balsa wood must be kiln dried. This is a two-week process that removes excess water until the moisture content of the timber is only 6%.

Commercially available balsa wood, typically found in model aeroplane kits, varies widely in weight. Balsa is occasionally found with densities as low as 64 kg/m³ (4 lb/ft³), and as high as 384 kg/m³ (24 lb/ft³). However, the commonly available density of commercial balsa for model aeroplanes will range between 96 kg/m³ (6 lb/ft³) and 288 kg/m³ (18 lb/ft³). Medium, or

---

4.3 A hollow column may be "stiffened" by internal pressure and then become capable of carrying a larger load. There are many examples of this type of action in nature.

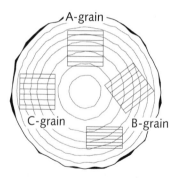

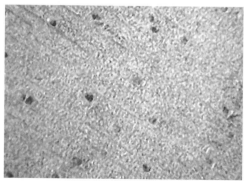

Figure 4.2(a) Various cuts of balsa from a balsa log. A–grain is called "tangent cut", C–grain is "quarter cut", and B–grain is "random cut"

Figure 4.2(b) Typical B-grain structure of random-cut balsa. The majority of the section shows the relatively open, cell-like structure of the timber. The solid, darker-coloured components are lignin, the gluelike polymer found in all timbers. (approx. 10x magnification)

average weight, balsa is the most readily available, ranging between 128 kg (8 lb) to 192 kg (12 lb). Relatively rare "contest grade" balsa, at 88 kg (5.5 lb) or less, is costly and difficult to obtain.

When selecting balsa sheets for a model, it is important to consider the way the grain runs through the sheet as well as the weight of the sheet. The grain direction controls the rigidity or flexibility of a balsa sheet more so than the density of the timber. For example, if the sheet is cut from the log so that the tree's annular rings run across the thickness of the sheet (A-grain, *tangent cut*), then the sheet will be fairly flexible edge to edge. In fact, after soaking in water some tangent cut sheets may be rolled into a tube shape without splitting. If, on the other hand, the sheet is cut with the annular rings running through the thickness of the sheet (C-grain, *quarter grain*), the sheet will be very stiff edge to edge and cannot be bent without splitting. When the grain direction is less clearly defined (B-grain, *random cut*), the most commonly available cut, the sheet will generally behave with intermediate properties between A- and C-grain. An essential rule of balsa model building is to carefully examine the grain structure so as to take best advantage of the timber's special characteristics. C-grain sheet balsa has a beautiful mottled appearance. It is very stiff across the sheet and splits easily. When used properly, it helps to build the lightest, strongest, most warp resistant, models. Figure 4.2 shows typical balsa grain structures milled from the log, with B-grain (*random cut*) the most commonly available type of balsa in model or hobbyist stores.

Mechanical Properties

Mechanical properties of balsa must be considered in the context of timbers generally. Balsa wood, in common with all timbers is an orthotropic material. It has unique and independent properties in the directions of three

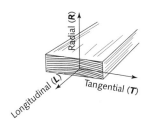

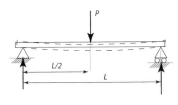

Figure 4.3 Three principal axes of wood with respect to grain directions and growth rings

Figure 4.4 Standard bending strength test for modulus of rupture in timber

mutually perpendicular axes, defined by the growth ring directions as shown in Figure 4.3.

Twelve constants (none of these are independent) are needed to fully describe the elastic behaviour of wood. These are three *elastic moduli*[4.4] E_i, three *moduli of rigidity* G_i and six *Poisson's ratios* μ_{ij}. Elastic moduli and Poisson's ratios are related by expressions of the form:

$$\mu_{ij} / E_i = \mu_{ji} / E_j; \mid i \neq j; \mid i, j = L, R, T.$$

Strength properties of wood are commonly obtained from static bending tests. The relative elastic and shear moduli are usually obtained from static compression tests. The *modulus of rupture* is a measure of maximum bending moment the specimen is able to withstand. Figure 4.4 shows the type of static bend test performed on a specimen of depth d and width b to failure under load P. The *modulus of rupture* is the extreme fibre stress at failure and is found from the relation $\sigma_B = 3PL/2bd^2$.

It is instructive to examine and compare these mechanical properties of balsa with other timbers. Table 4.1(a) lists some comparative elastic ratios and Table 4.1(b) lists comparative Poisson's ratios for the same set of commonly occurring timbers.

Engineers use great varieties of materials in the construction of structures and machinery. The size of components ultimately used in a construction is often dictated by the relative mechanical properties of the materials being used. When making use of unconventional engineering materials for MaT projects it is instructive to compare mechanical properties of these unconventional materials to those of commonly used engineering materials. Table 4.2 lists comparative mechanical properties for some materials.

It is further instructive to compare mechanical parameters (combinations of mechanical properties) often used in the design of structures and machines. One example is the ratio of rupture tensile strength to specific gravity (this parameter may be loosely identified as strength to weight ratio for beams). Using Table 4.2, we find the unexpected comparative values of this parameter for steel (39 average), aluminium (average 30) and balsa

4.4 Terminology is defined and explained in *"Glossary of Commonly Used Design Terms"* at the beginning of this book.

Table 4.1(a) Elastic ratios for some common timbers (12% moisture content)[4.5]

Species	E_T/E_L	E_R/E_L	G_{LR}/E_L	G_{LT}/E_L	G_{RT}/E_L
Balsa	0.015	0.046	0.54	0.037	0.005
Mahogany, African	0.050	0.111	0.088	0.059	0.021
Oak, white	0.072	0.163	0.086	—	—
Cedar, western red	0.055	0.081	0.087	0.086	0.005
Pine, longleaf	0.055	0.102	0.071	0.060	0.012

Table 4.1(b) Poisson's ratios for some common timbers (12% moisture content)

Species	μ_{LR}	μ_{LT}	μ_{RT}	μ_{TR}	μ_{RL}	μ_{TL}
Balsa	0.229	0.488	0.665	0.231	0.018	0.009
Mahogany, African	0.297	0.641	0.604	0.264	0.033	0.032
Oak, white	0.369	0.428	0.618	0.300	0.074	0.036
Cedar, western red	0.378	0.296	0.484	0.403	—	—
Pine, longleaf	0.332	0.365	0.384	0.342	—	—

(135). Although these figures might imply that balsa beams could perform better than steel or aluminium beams in some applications, there are many other mechanical properties that will influence the final choice of construction material. Nevertheless, this example suggests that unconventional materials, used for MaT project construction, should not be easily dismissed as useful only in model building exercises.

4.2 Newsprint

Another versatile modeling material is the ubiquitous newspaper page, commonly taken from a broadsheet newspaper such as the *New York Times*, the *London Guardian* or the *Melbourne Age*. A double page of the *Melbourne Age* newspaper measures approximately 600 x 810 mm (24 x 32 inch) with a thickness of 0.0625 mm. The size of this double sheet varies slightly from the standard US newsprint (24 x 36 inch), probably due to some amount of

4.5 Source: *Wood Handbook* (1999): Chapter 4, Green et al. *Mechanical properties of wood.*

Table 4.2

Comparative mechanical properties for some materials (timbers at 12% moisture)[4.6]

Material	Specific gravity (SG)	σ_B MPa ($10^6 N/m^2$)	E_L (GPa)	Compression parallel to grain σ_C (MPa)	σ_B/SG
Balsa	.16	21.6	3.4	14.9	135
Mahogany, African	.42	73.8	9.7	44.5	175.5
Oak, white	.64	71	9.4	24.3	111
Cedar, western red	.32	51.7	7.7	31.4	162
Pine, longleaf	.59	100	13.7	58.4	170
Mild steel (low carbon steel)	7.7	250–350	210	—	32.5 – 46
Aluminium	2.7	50–114	7	—	18.5 – 42
Magnesium	1.74	90–220	44	—	52 – 126
Kevlar	1.38	338	125 –130	—	245
Titanium	4.51	345 to 552	110– 125	—	79 –122
Carbon fibre	1.8–2.1	1193 – 2700	200– 400		596 – 1350

trimming. Consequently, whenever a MaT project uses newsprint as a material of construction, the size of the page should be identified to maintain consistency in project specifications. Although newsprint is a part of the much larger general paper family, it is designed to be cheap and to absorb printing ink readily. These properties are imparted during the pulping process used for making newsprint material. Moreover, it is also softer and easier to recycle than other forms of paper. Its abundance and ease of availability makes newsprint a useful model making material. To place this material in the context of the paper family it is useful to review a brief history of paper and papermaking.

Paper consists essentially of cellulose fibres, which are the main component of pulp, the raw material from which paper is made. The individual fibres are present in a network, as can be seen by looking at the torn edge of a piece of paper. These fibres occur in nature and are made up mainly of

4.6 Sources: Wood Handbook, Ibid; Samuel and Weir (1999).

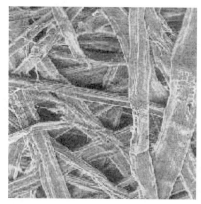

Figure 4.5 Typical fibre distribution in newsprint (approx. 100x magnification)

cellulose, the basic constituent of wood, from which paper pulp is derived initially. In the finished paper, each cellulose fibre is bonded to its adjacent fibres by large numbers of hydrogen bonds. These chemical bonds are weaker than the chemical ionic and covalent bonds that hold most materials in the world together, but they provide the main strength properties of paper. Mechanical entanglement of the fibres makes only a minor contribution to holding fibres together. Figure 4.5 is a photomicrograph of typical newsprint. Paper properties differ widely, according to their grade, and the properties of individual grades are achieved by appropriately influencing the components of the raw materials of paper during the relevant pulp and paper manufacturing stages.

Paper was invented in China, as a substitute writing surface to silk, around 105 A.D. Mechanical papermaking is due mainly to a papermaking machine invented by Frenchman, Nicholas Louis Robert in 1798, developed in England by Brian Donkin for Henry and Sealy Fourdrinier, but not placed into operation until 1804. The *Fourdrinier machine* and the *cylinder machine* are normally employed in the manufacture of all grades of paper and board. In the Fourdrinier machine the wet pulp, in a concentration of approximately 0.2% to 1%, depending on paper grade, is fed from a container (the *headbox*) through a slit onto a moving wire belt (the *Fourdrinier wire*). A thick wet "feltlike" sheet of paper is produced in this process. The wet sheet is stripped off the wire and then passed through several water removal and drying stages before it is consolidated into rolls of commercial grade paper.

The *cylinder machine* has one or more wire mesh cylinders or *moulds* that are partially immersed and rotated in vats containing a dilute pulp suspension. The pulp fibres cling to the wire and are formed into sheets on the cylinders (similar to the feltlike wet sheet formed in the Fourdrinier process) as the water drains through and passes out at the ends of the cylin-

Figure 4.6 An early Fourdrinier machine, with wet end and head box on the left

ders. Water removal, drying and finishing follow in a manner similar to that of the Fourdrinier process. Figure 4.6 shows an early type of Fourdrinier machine.

Although this narrative has been partly motivated by the possible use of newsprint as a MaT project construction material, it is useful and instructive to enlarge the discussion to embrace all forms of paper. No doubt, ingenious MaT project designers will find ways of using paper other than newsprint in future MaT projects. Overall paper quality is broadly defined by *size*, *grammage* and *caliper*. Size is dealt with more fully below.

> *Grammage* is a measure based on the mass of a *ream* of a specific basic size in the size ranges of paper. This ream-mass is then converted to a measure of grams per square meter or *gsm* (whence the term grammage). Typically:

- 32 lb US newsprint (24 x 36 inch) has a grammage of 52 gsm, and

- 12 lb bond paper (17 x 22 inch) has a grammage of 45 gsm.

> *Caliper* is the thickness of paper expressed as thousands of an inch (1 mm = 40 x 10^{-3} inch). Hence, the *Melbourne Age* newspaper has

Table 4.3 Paper sizes in current usage[4.7]
Formerly, all individual sheet sizes had their own name, which was specific to a particular country or region. The following names are currently used in North America

Name	Dimensions (inches)	Dimensions (mm)
Folio	17 x 22	432 x 559
Letter	8.5 x 11	216 x 279
Legal	8.5 x 14	216 x 356
Monarch	7.25 x 10.5	184 x 267
Medium	18 x 23	457 x 584
ANSI (American National Standards Institute) sheet sizes		
A	8.5 x 11	
B	11 x 17	
C	17 x 22	
D	22 x 34	
E	34 x 44	
UK named sizes—common before metrication, and still used		
Quarto	8 x 10	203 x 254
Foolscap	8 x 13	203 x 330
Royal	20 x 25	508 x 635
Double crown	20 x 30	508 x 762
Double cap	17 x 27	432 x 686
Large post	16.5 x 21	419 x 533

4.7 Source: http://www.paperloop.com.

Table 4.3 Paper sizes in current usage (continued)

A Series – mm/(inch)		B Series – mm/(inch)	
4A0	1,682 x 2,378	4B0	2,000 x 2,828
2A0	1,189 x 1,682	2B0	1,414 x 2,000
A0	841 x 1,189	B0	1,000 x 1,414
	(33.11 x 46.81)		(39.37 x 55.67)
A1	594 x 841	B1	707 x 1,000
	(23.39 x 33.11)		(27.83 x 39.37)
A2	420 x 594	B2	500 x 707
	(16.54 x 23.39)		(19.69 x 27.83 in)
A3	297 x 420	B3	353 x 500
	(11.69 x 16.54)		(13.90 x 19.69)
A4	210 x 297	B4	250 x 353
	(8.27 x 11.69)		(9.84 x 13.90)
A5	148 x 210	B5	176 x 250
	(5.83 x 8.27)		(6.93 x 9.84)
A6	105 x 148	B6	125 x 176
A7	74 x 105	B7	88 x 125
A8	52 x 74	B8	62 x 88
A9	37 x 52	B9	44 x 62
A10	26 x 37	B10	31 x 44
C Series: for envelopes and folders that are to contain A Series material.			
C4	229 x 324	A4 unfolded	
C5	162 x 229	A4 folded in half	
C6	114 x 162	A4 folded in quarters	
DL	110 x 220	A4 folded in thirds	

International Sheet Size Series: *There are five main international series, although minor deviations from these sizes are sometimes encountered.*
A Series: *This is an ISO series of standard trimmed sheet sizes for printed matter and stationery. Every size has its sheet length and width in the same ratio of $1:\sqrt{2}$ (i.e., 1 to 1.414). The basic size, A0, is 841 x 1,189 mm. Each subsequent size is obtained by either doubling or halving the longer dimension.*
⌊**B Series**: *An ISO standard of trimmed sheet sizes, with the same ratio of dimensions as the A Series. It is intended for posters and wall charts, etc., and fills the gap between adjacent sizes in the A Series.*

a *caliper* of 1.56×10^{-3} inch. The need to define paper quality is due to the significant growth of mechanization in the paper handling industry. Paper feeding and delivery are the most complex mechanical parts of the printing press, therefore consistent paper quality is mandatory for machines handling paper at some speed.

In the early part of the twentieth century, before World War I, many industrialised countries discussed the advantages of standardising paper sizes to permit large-scale mechanised handling of mail. Most countries accepted the *International Standards Association's* recommended standard sizes in 1925. The United States held a joint meeting of manufacturers, distributors and users resulting in the *Simplified Practice Recommendation R22*, effective June 1933. This recommendation identified the existing sizes in most frequent use, but did not add any new ones. Table 4.3 lists the paper sizes in common usage as well as the current *International Standard* sizes.

Ream is one of the oldest terms in the paper industry, and is the term used universally for a quantity of paper. Its origin may be traced to the Arabic term *rizmah*, meaning a bundle (of paper or cloth), and *rasma*, meaning to collect into a bundle. The actual quantity of paper in a ream gradually became fixed through the centuries and probably has some relationship to the amount of paper a *vat-man* (the person lifting sheets of paper from a vat of pulp on a framed wire or cotton mesh) could make in a day. Variations still exist today, according to the type of paper sought, but in general use a 500-sheet ream of paper tends to be standard.

The most likely reason we use rectangular paper dates back to the early days of paper production. Until the beginning of the nineteenth century, paper was made by hand, dipping a mould into a large vat containing beaten pulp in suspension. Early papermakers found it much easier to control rectangular moulds than square ones. The rectangular size tradition continued after the process was mechanised in the late nineteenth century, and mechanical blades were used to cut the rolls of paper for the commercial market. Tradition has kept the shape rectangular, and metric size ratios (width to length) are held in the ratio $1:\sqrt{2}$.

Newsprint is a generic term applied to the type of paper used for printing newspapers. The paper is made in basis weights of 30 to 35 lb, the average weight of 500 sheets of 24 x 36inch paper (the *SI* equivalents are 61 x 915 weighing 1360 to 1588 gm, which converts to 48.7 to 56.9 gsm). The most commonly produced is 32 lb basis weight (which converts to 52 gsm). This 52 gsm paper has proven to be the most satisfactory practical compromise between cost and printability.

Pulping is the process of turning finely chopped timber into the mix of water and cellulose fibres, that represent the raw material in paper-making machinery. Broadly speaking, pulp is produced either by mechanical means in a beater, or chemically in a digester. Digesters cook (digest) wood chips in an alkaline solution for several hours, during which time the chemicals attack the lignin that binds the cellulose fibres together in the wood. The beater is, in essence, a type of threshing machine, used for mechanically separating the cellulose fibres in the wood.

Newsprint is produced at very high machine speeds, which causes the fibres of the sheet to be directional. However, the fact that the majority of the fibres are oriented in one direction is actually an advantage from the point of view of printing, because it helps the paper take the strain of the high-speed press used for printing newspapers. From the point of view of newspaper production, this is the most important characteristic in newsprint. The usual raw material for newsprint is 75% to 85% mechanical pulp, and 15% to 25% unbleached or semi-bleached chemical pulp. The mechanical pulp adds the desired properties of high opacity, smooth surface, and high oil (ink) absorption, and the chemical pulp adds the necessary strength required to run the paper through fast rotary presses without breaking.

Mechanical Properties of Paper

Important mechanical properties of paper are tensile strength, *collar crushing* strength (for stackability as packaging material) and *bursting strength* (also a factor related to load bearing capability in certain types of packaging). The tensile strength is often expressed as "breaking-length", *LB*, that represents the length of a particular type of paper that can hang vertically under its own weight without breaking. Breaking-length is a useful comparative measure of strength for thin sheets of material. It is measured by considering a roll of sheet material hanging under its own weight at failure, as shown in Figure 4.7. Although this is an indirect method of expression for tensile strength it permits comparison of a range of materials. As an example, Table 4.4 gives breaking lengths of various materials.

Collar crushing strength is measured by loading a short column of paper axially to failure. *Bursting strength* is measured by clamping a sheet of paper to a rubber diaphragm which is inflated until the paper sheet bursts (*Mullen's test*). Due to the wide range of paper types available, tables of published strength properties tend to be proprietary information to the paper manufacturers. In general, it is useful to establish, by direct testing, the tensile, crushing and column failure properties for the paper specified in the MaT project. Structural test results for relevant paper properties are given in Section 4.4 below.

Table 4.4 Breaking length of various materials[4.8]	
Material	Breaking length LB km
Eastern white pine	22.7
Newsprint	2 to 5
Paper from bleached soft wood	8 to 10
Steel	4.5
Aluminium	3.4
Graphite	37

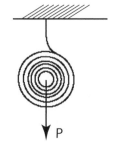

Figure 4.7 Breaking length measurement (schematic)

4.8 Source: http://natural-resources.ncsu.edu: 8100.

4.3 Polymers, Composites and Other Strange Structural Materials

A wide variety of composite materials may be specified for MaT projects. The easiest to procure are the polyurethane foams and *foamcore*, a product formed by sandwiching a polyurethane core between two sheets of paperboard. Rigid polyurethane foam sheets are also readily available from manufacturers of insulation material. It is easily worked with a utility knife or razor blade. Foamcore is often used for architectural model building and is readily available from hobbyists or art suppliers.

Rigid Polyurethane Foam and Foamcore[4.9]

Rigid polyurethane foam is manufactured in a wide range of densities from 35 to 250 kg/m³. It is made from a two-part mix of chemicals with a CO_2 *blowing agent*. CO_2 gas is trapped in the foam as a series of small bubbles and this bubble structure makes the foam light and resistant to heat transmission. Figure 4.8 shows typical foam structures for medium and high-density polyurethane foam, both at approximately $10\times$-magnification. It is clear from these images that the lower-density bubble structure is less regular and includes some larger bubbles among the smaller ones, whereas the more regular structure of the higher-density urethane appears to be formed from a regular dispersion of smaller bubbles.

Foamcore (occasionally referred to as *Styrofoam*, when the core is styrene) is one of several other easily worked composites commonly used in architectural model building. Hence it lends itself readily to MaT model construction. Foamcore, as the name implies, has a central core of styrene or urethane foam sandwiched between two cardboard skins. The cardboard skin material (a paper product) is very much stronger in tension than the foam. So, from a structural point of view, foamcore may be regarded as a type of *I-beam*, where the central material is simply a separator of two load bearing

Figure 4.8(a) High density foam structure   *Figure 4.8(b) Medium density foam structure*

4.9 The samples of urethane tested in this chapter were kindly provided by Australian Urethane and Styrene Pty Ltd.

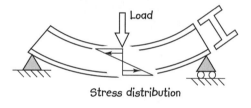

Load

Stress distribution

Figure 4.9(a) Section through 8 mm foamcore at approximately 6 x magnification

Figure 4.9(b) Idealised model of bending for a composite beam (foamcore) where the core material is significantly weaker than the outside "skin"

skins of material. In addition, the skins provide some protection against local surface damage as well as a smooth appearance. Figure 4.9(a) shows a section through foamcore. Figure 4.9(b) is an idealised view of the bending behaviour of composite beams such as foamcore. The weaker core material is reduced to a thin web separating the two flanges of the idealised beam.[4.10] By modelling the composite beam in this way we are able to analyse the stresses in it as though it were made of a single material. Where composite materials are used for MaT projects, simple testing is always recommended rather than analysis. This type of experimental evaluation of behaviour is especially important in the case of *foamcore*. Since the core material is not only much weaker in tension than the cardboard "skin", but also much weaker in local compression. As a result, the most common mode of failure in *foamcore* is local buckling of the cardboard skin. When used in bending applications, the upper surface of the *foamcore* beam is subject to compressive stresses that will result in just such local buckling.

Bamboo Skewers

For some considerable time now, bamboo has been regarded as an ecologically friendly material to replace timber or (as in Figure 4.10) steel in some situations. As a member of the grass family, bamboo grows rapidly and is quite abundant. It is stronger than many timbers and it may be laminated into slabs of material to be used instead of other natural wood products. In its application for timber floor replacement, published performance data is available from several bamboo floor suppliers.[4.11] Table 4.5 gives some published data for bamboo flooring, reconstituted from natural bamboo stalks, by flattening and laminating the outer hard shell of bamboo tubing. In ladders, scaffolding or fencing, bamboo is twice as stable as oak, and harder than walnut or teak. After at least three years of growth, the bamboo's hollow round tubes are sliced into strips, boiled to remove the starch, then dried and laminated into solid boards.

4.10 See also Timoshenko (1981).
4.11 See also Arce-Villalobos (1993); Janssen (1995).

Figure 4.10 Bamboo scaffolding on a building in Hong Kong

Table 4.5 Data for bamboo flooring material[4.12]	
Physical property	Value
Elastic modulus in bending	6.5 GPa
Modulus of rupture in bending	93.9 MPa
Compressive strength (parallel to grain)	52.1 MPa
Compressive strength (perpendicular to grain)	17.7 MPa
Load to compress to 50% strain	116.7 MPa
Coefficient of friction (on metal)	0.71

Although bamboo is a very important structural material, it has many other features that make it of practical importance to technology generally. It was the needle in Alexander Graham Bell's first phonograph and Thomas Edison used it in 1882 for the world's first incandescent light bulb filament. It can grow half a metre per day and reach a height of 30 m. A bamboo stand generates more oxygen than an equivalent stand of trees. It survived the Hiroshima atomic blast closer to ground zero than any other lifeform and, as a result, it is regarded as a nuclear-tough.

Some of its structural applications include bamboo bicycle frames in the late nineteenth century and bridge construction. In China, a bamboo suspension bridge is 220 m long, 3 m wide and rests entirely on bamboo cables fastened over the water. It is entirely free of metal connectors. Bamboo compares very favourably with steel as a construction material for tensile structures (193 MPa compared to 210 MPa for structural grade mild steel); on strength-to-weight ratio in tensile structures it is nearly eight times better than steel.

Drinking Straws, Spaghetti

The range of materials available for MaT project construction is limited only by our imagination. In this text only the most commonly used options are considered. Spaghetti, a readily available commodity in most supermarkets, has been widely used in local and national "build-and-break" competitions. It is easily glued with balsa wood glue or quick-setting polyester based glues. Moreover, in the dry state, it is quite strong and flexible.

4.12 Source: Canada Bamboo: www.floors.ca

Interestingly, most of the competitions using spaghetti as a structural material explore only its capacity to carry a load. Yet there are many opportunities for spaghetti-based projects that also test flexure and structural stability. (see, for example, the egg support structure of Figure 2.2). Drinking straws (paper or plastic) and bamboo skewers (as used in *kebab*) offer further opportunities to explore unconventional engineering structures. The exercises at the end of this section list some of these opportunities.

Cotton Thread

Thread is wonderful! It has to be wonderful, since you expect it to be capable of travelling at speeds of 200 miles per hour, to survive temperatures of 800 degrees Fahrenheit, to resist attempts to break at a rate of 180 times per second, to withstand exposure to acids and alkalis, to scoff at the effect of millions of abrasion strokes, to be twisted, bent, and coiled into intricate configurations, and to sell for only a few hundredths of a cent per meter. The thread industry is proud that we can and do make such a wonderful product.

American Textile Manufacturers Institute

Cotton thread is available in a range of thicknesses and strengths for mercers and tailoring suppliers. It may be used for strengthening joints (as has been done in early model aeroplane construction), or for tensile supporting structures (refer to Chapter 8, tensegrity structures). The history of cotton thread predates paper and it has been used in Pakistan for weaving cloth for over 5000 years. In its raw state, cotton thread is made from cotton yarn, each thread formed from several yarns. Thread size has metamorphosed through various designations into, the now internationally accepted *Tex* measure. This is the weight of 1000 m length of a specific thread. Normal sewing thread is of basic breaking strength. Specialty threads, including polymer mixes (such as *nylon* or *kevlar*) an substantially increase the strength of a given Tex number thread. *Mercerizing* (a chemical treatment) also improves thread breaking strength substantially. Table 4.6 gives the basic single end breaking strengths of various size cotton threads.

Table 4.6 Thread types and breaking strengths[4.13]	Tex size	Breaking strength	
		lb force	kg force
Core cotton wrapped (D-core)	24	2.3	1.04
	50	4.6	2.09
	105	7.5	3.4
	150	15.1	6.9
CF Twisted multifilament (Anefil Nylon Poly)	30	4.7	2.1
	45	7.4	3.4
	90	14.5	6.6
	135	23.5	10.7
	270	44.2	20.0
	400	73.3	33.3

4.13 Source: American & Efird Inc.

4.4 Establishing Mechanical Properties for Materials Used in MaT Model Construction

Published mechanical properties for polyurethane foam provide compressive stresses at 10% deformation ranging from 100 kPa to 175 kPa. Information about other structural properties, particularly relating to some of the less conventional structural materials discussed above, is difficult to find. All mechanical properties needed for MaT project construction using these unconventional materials have been found from direct testing by the author.

Most material properties vary sufficiently to make it necessary, and often instructive, to perform simple tests on specific materials to determine actual properties rather than simply accepting published data. The following tests are simple to perform with minimal test equipment. The tests have been constructed to make it possible to perform them with materials and skills found in most private homes. These home-grown simple tests are not likely to be as accurate as those performed in a materials laboratory, however, if applied with some care, the results can closely reflect the properties of actual materials used in MaT model construction. In addition, some MaT model construction materials, such as newsprint or spaghetti, have no readily available published mechanical properties.

In these structural tests consistent units of measurement reflect the available measuring instruments used. Distances and sizes are given as measured with a metre rule converted to millimetres. Loads are shown as grams measured on a kitchen scale.

Testing Balsa Wood

(a) Test for Modulus of Rupture

Figure 4.11 shows a simple setup used for these types of tests and Figure 4.12 shows the load deflection curve found for the balsa sample. In this test a piece of timber board is fitted with two bolts at (as precisely as possible) 300 mm centres. These bolts provide (approximately) frictionless simple supports to the test piece. A carpenter's rule is set up vertically and central to the two support bolts. In the test, shown in Figure 4.11, the test piece is balsa wood. A golf tee, attached to a scale pan by 40 kg fishing line, provides the means of loading the beam. The point of the tee permits recording deflection as weights are added to the scale pan. The weights can be measured on a kitchen scale.

Based on the load-deflection behaviour the following can be established for the sample balsa tested:

Section proportions = 20.5 mm wide x 8.3 mm deep

Load at rupture = 6170 gm

Span = 300 mm

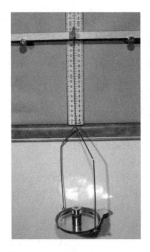

Balsa beam rupture test

Load gm

600

400

200

0   2   4   6   8   10

Deflection mm

Figure 4.11 Simple test for modulus of rupture

Figure 4.12 Load deflection curve for balsa beam

Hence the modulus of rupture

$$\sigma_B = (3 \times 6170 \times 9.81 \times 0.3)/(2 \times 20.5 \times 8.32 \times 10^{-9})$$

$$= 19.3 \text{ MPa}$$

Weighing a piece of this specific sample of balsa yielded the approximate specific gravity of 0.1. Because this specific sample is much lighter than the sample data offered in Table 4.2, as expected the modulus of rupture is also significantly reduced (from 21.6 MPa to 19.3 MPa). Clearly, even with experimental errors introduced into our simple test, the result is likely to be a closer indication of the real strength of the sample material than the published data.

In the test for modulus of elasticity, the beam was preloaded to remove any kinks and twists from the fishing line supporting the scale pan. A further load of 743 gm was then added to the scale pan and this increased load resulted in an increase in beam deflection of 1.2 mm. Using the formula for the elastic deflection of a centrally loaded simply supported beam we find:[4.14]

$$\delta_{MAX} = PL^3/(48EI), \text{ which when rearranged provides}$$

$$E = PL^3/(48\delta_{MAX}I).$$

This results in

$$E = (0.734 \times 9.81 \times 0.33)/(48 \times 0.0012 \times 9.8 \times 10^{-10})$$

$$= 3,440 \text{ MPa}.$$

4.14 A brief introduction to elementary beam theory is given in Appendix 1.

Figure 4.13(b) Cylindrical collar crushing specimen

Figure 4.14 Paper column construction

Figure 4.13(a) Tensile specimen

Figure 4.13(c) Pleated crushing test specimen

Figure 4.15 Column base detail

Some Notes on Experimental Procedure

- Balsa wood is soft and care is needed to prevent local damage of the timber during loading. For this reason it is desirable to provide a loading pad to distribute the load under the loading device. In the test described the loading device was a wooden golf tee 3.5 mm in diameter and a small piece of balsa under the tee was used as the loading pad.

- To evaluate deflection, it is good experimental practice to load the beam to some initial load to ensure both bedding in of the supports and elimination of minor kinks and twists in the loading wire (see Figure 4.11).

Testing Newsprint

As indicated earlier, due to its mode of manufacture, mechanical properties of newsprint tend to vary substantially with direction of loading relative to machine direction of manufacture. Tensile testing was performed in both machine-direction and cross-direction. Collar crushing and column tests were only performed in the cross direction, as this property is not seriously influenced by manufacturing (or long fiber) direction.

Figure 4.13 shows the various test procedures used in testing the mechanical properties of newsprint.

Figure 4.13(a) shows the simple tensile test specimen (6.88 x 0.0625 mm in the central thin section) used. The photo in the figure shows a machine direction sample under test. A specimen of newsprint was cut and held by bulldog clips at each end. An eraser was inserted into the jaws of the bulldog clips, together with the ends of the test specimen, to ensure sufficient friction in the grip during loading. Other friction material, such as soft rubber, may be used in the tensile test, but this aspect of the test is left open to personal experimentation. A small keyring attached one clip to a bolt in the board used for the balsa tests and the legs of the other clip were attached to a scale pan.

Figures 4.13(b) and 4.13(c) show the specimens used for collar crushing tests. The simple cylinder was made from the same length of paper and rolled to approximately the same mean diameter as that of the pleated specimen of Figure 4.13(c) (approximately 85 mm diameter). Each specimen was 130 mm long. Crushing loads were applied to these specimens while mounted between two sheets of foamcore directly on a kitchen scale. Table 4.7 lists the results of these tests.

Long columns of paper were also tested for buckling. These columns were constructed from approximately one-half of a broadsheet [4.15] (8 gm) with their axes in the machine direction. The sheet was ironed to remove kinks and any moisture that might weaken it. The paper was rolled onto a timber dowel and the rolling continued until the paper roll was quite consolidated. Then the timber dowel was removed and a thin metal rod was used to further decrease the column diameter and consolidate the paper by continuous rolling. In this way a consistent well-consolidated column of paper 8.9 mm outside diameter was constructed. Figure 4.14 shows the method of construction.

Once constructed, paper models need special care in manufacturing column bases and various joints for the eventual load bearing structure. Failure in these types of structures inevitably occurs at poorly constructed locations, such as column ends (local crushing and destabilized structure behaviour), and at weak or poorly located joints. Considerable experimentation is needed for exploring alternative arrangements. In particular, paper column bases may be protected against local crushing by a substantial increase of the column diameter or thickness near the base. Lateral stabilisation can be achieved by forming splayed connection at the column base as indicated in Figure 4.15.

Testing of Other Unconventional Structural Materials

Axial load testing of drinking straws and spaghetti may be performed with a relatively simple structure. Figures 4.16 and 4.17 show the style of test used in these axial loading experiments. Care is needed to permit some lateral

---

4.15 Most common newsprint is made of 32 gsm paper, giving a broadsheet weight approximately 16 gm.

rotation of the column at the ends during loading. Unnecessary constraining of the ends of the column can yield false results. In the experiments reported here the ends of these straw and spaghetti columns were located in miniature plastic cups (plastic covers of screw holes in household electrical fittings). In the case of the axial load test for bamboo skewers (Figure 4.16) the buckling load may be read directly from the scale on which the test is being performed.

With cotton (or other mixed material) threads,

 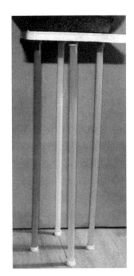

Figure 4.16 Axial loading bamboo skewers

Figure 4.17 Axial loading drinking straws

the information in Table 4.6 provides order of magnitude values only. If thread strength is a key design variable, simple load testing should be used. The same "test apparatus" used in all the other material tests is certainly appropriate for such a test. In the case of nylon fishing wire, kevlar, or other high-strength threads, the manufacturer's specification should be sought.

Some Additional Notes on Testing and Failure Behaviour

- *Balsa wood*
  Samples failed quite suddenly on the tensile surface of the beam. Care was needed to ensure that loading was well distributed, because local stress concentration due to point loading severely diminished beam performance.

- *Urethane foam*
  These samples also exhibited sudden failure. The failure surface has a brittle appearance although the material seems quite flexible before it fails.

- *Paper short columns*
  These failed by local crushing and the failure load is expressed as load per unit length of the circumference.

- *Bamboo skewers*
  These failed by "flexing out" from the supports. They did not break, although Table 4.5 lists published rupture data for bamboo.

Table 4.7 Mechanical properties for some unconventional engineering materials[4.16]

Material (all dimensions in mm units)	Specific gravity	σ_B MPa $(10^6 \, N/m^2)$	E MPa
Bending and tensile loading tests			
Balsa wood (mixed grain)	0.1	19.3	3,440
Urethane foam (low-density)	0.35	0.23	3 – 4
Urethane foam (high-density)	0.9	0.58	11
Foamcore (5.4 thick)	0.115	3.6	687
Newsprint – machine direction	43 gsm[1]	32	—
Newsprint – cross direction	43 gsm[1]	12	—
Bamboo skewers (kebab) bending	0.81	—	11,300
Spaghetti (dry, 1.8 dia.)	1.43	20.4	3,500
Axial loading tests			
Material (all dimensions in mm units)	Specific gravity	Failure load	E MPa
Newsprint crushing test (plain cylinder)	43 gsm[1]	11 N/m	—
Newsprint crushing test (pleated cylinder)	43 gsm[1]	15 N/m	—
Newsprint tubing (8.9 dia. x 1.125 thick x 459 long)	43 gsm[1]	$P_{cr} = 49$ N	5,800
Bamboo (kebab – 2.6 dia. x 250 long)	0.81	$P_{cr} = 3.9$ N	11,100
Spaghetti (dry, 1.8 dia. x 250 long)	1.43	$P_{cr} = 0.29$ N	3,300
Plastic soda straws (5.25 dia. x 0.15 thick x 165 long	$0.11(1.2^2)$	$P_{cr} = 2.94$ N	1,430

[1] For newsprint specific gravity (ratio of the mass of a given volume of material to the mass of the same volume of water) is not as useful as the gsm (grams/m²) unit.
[2] Specific gravity of the solid plastic part of the soda straw.

4.5 Adhesives[4.17]

The range of available adhesives is considerable. The topic of adhesive design is broad and, for those interested in the chemistry, physics and mechanics of adhesion, there are international research organizations and

4.16 Source: direct testing by author
4.17 For technical properties and chemistry see Pocius (2002); Adhesives: Structural Adhesives Directory and Databook (1996); Ahesives: Wood Adhesives (1995); Pizzi and Mittal (2003); Composites Bonding (1994).

journals available. Because joints are likely to be critical design points in a structure, considerable care is needed in selecting a suitable adhesive for joining structural members. A brief list of available adhesive types is given below. However, in designing MaT projects, one must be guided by what is readily available from local suppliers. The choice need not be too restrictive, inasmuch as adhesive choice and joint design might provide some creative freedom for designers. For paper-based projects several different types of adhesive tape may be specified. The allowable sizes and lengths to be used in the project should be carefully limited, because these tapes are significantly stronger in tension than paper. In yet other structural MaT projects the use of adhesives might be excluded completely.

Adhesive Type Definitions

- *Two-component (room-temperature cured)* adhesives (including epoxy, urethane, polyester, acrylic and silicone): These are adhesive systems that require, just prior to application, the addition of a second component. The two parts must be measured and thoroughly mixed. The second component is commonly referred to as the *curing agent, hardener, catalyst, activator* or *accelerator*. The two components are also sometimes described as *part A* (resin) and *part B* (curing agent). These systems, once mixed, are designed to cure at room temperature, although the cure time may be accelerated with heat.

- *Two-component (high-temperature cured)* adhesives (including epoxy, polyester and silicone): These are adhesive systems with two parts similar to room-temperature curing adhesive mixes, with special curing requirements. These systems, once mixed, are designed to cure only at elevated temperatures (generally 80°C minimum). Cure temperatures and schedules are often provided to permit a variety of cured characteristics.

- *Single-component (room-temperature or heat cured)* adhesives (including epoxy, urethane and silicone): In these the catalyst or hardener is an integral part of the system and requires no mixing. These adhesives may have limited shelf life, and/or special storage conditions, such as refrigeration, may be required. Curing can be activated by introducing heat (epoxies), oxygen (anaerobic) or moisture (silicones).

- *Single-component (ultra-violet cured)* adhesives: High-power UV light sources can cure these adhesives very quickly when needed for high-speed assembly work or applications that require a fast cure.

- *Single-component (solvent)* adhesives: These contain some form of solvent, including organic and aqueous (water-based) solvents within the adhesive system that is usually driven off during the cure cycle. Although they can form excellent bonds, they are often plagued by flammability issues or vapor/air quality problems. One advantage of this type of system is that their generally lower viscosities can aid the application process.

- *Hot melt adhesives*: These are fast-setting adhesives based on thermoplastic polymers that are solid at room temperature and start to activate when heated above their softening point. Bonding is achieved when the adhesive cools. Hot melt adhesives are well suited for bonding porous materials, bcause they are generally used in applications at room temperature and tend to be highly viscous.

- *Cyanoacrylates*: These have the ability to polymerize very quickly at room temperature without a catalyst and are suitable for use with a variety of bonding surfaces. They have been widely used for thread locking applications, and sometimes are subject to hygroscopic degradation. They are limited in temperature capabilities and chemical resistance.

- *Contact adhesive systems*: These are solvent-based adhesives normally applied by spray, roller or brush. Coating both substrates with the adhesive and bringing the surfaces together using only contact pressure form the bond. One advantage of this type of adhesive is early bond strength upon contact. As with many other solvent-type adhesives, there can be problems with flammability and air quality. In many cases, this type of adhesive may be limited in temperature range as well as chemical resistance.

- *Anaerobic adhesives*: These are fast-setting adhesives that cure in the absence of oxygen. They need to be packaged in oxygen-penetrating containers and usually filled only partially to allow plenty of oxygen-rich headroom in the container. Curing occurs when the bond line cuts off the oxygen and thus close fitting surfaces are desirable.

- *Film adhesive systems* (including supported and unsupported film adhesives): These may be available as unsupported or supported by a film or fabric carrier. Many of these systems are in the form of partially cured adhesives that have been applied to a carrier. The adhesive may be cut to shape then placed into position on a substrate and the release cover removed. The substrates are joined under pressure and heated to cure. (A typical example is double-sided adhesive tape.)

- *Electrically conductive adhesives* (including epoxy and solvent systems): These adhesive systems are usually heavily filled (greater than 70%) with electrically conductive fillers (silver, copper, gold, nickel and carbon). Conductivities of 0.001 ohm/cm and less can be achieved with many of these systems. Typical applications include replacement for solder in electrical connections, connections to substrates that can't be soldered or where heat would be a problem.

4.6 Test Results for Mechanical Properties of MaT Project Construction Materials

The following pages present detailed evaluation of test results used in determining mechanical properties of various unconventional structural materials.

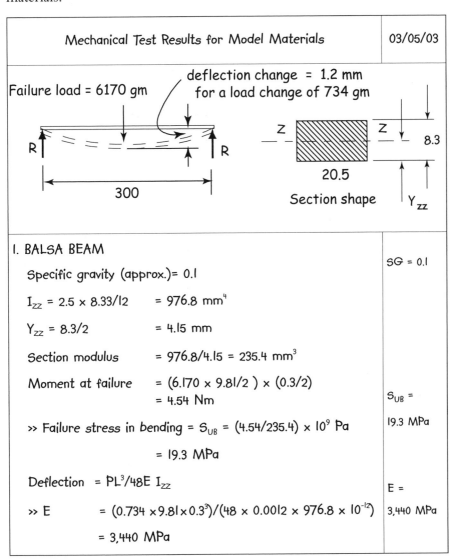

Mechanical Test Results for Model Materials	03/05/03

Failure load = 6170 gm

deflection change = 1.2 mm
for a load change of 734 gm

R ↑   R ↑   8.3

300   20.5

Section shape Y_{zz}

1. BALSA BEAM

SG = 0.1

Specific gravity (approx.)= 0.1

I_{zz} = 2.5 × 8.33/12   = 976.8 mm⁴

Y_{zz} = 8.3/2   = 4.15 mm

Section modulus   = 976.8/4.15 = 235.4 mm³

Moment at failure   = (6.170 × 9.81/2) × (0.3/2)
     = 4.54 Nm

S_{UB} =

>> Failure stress in bending = S_{UB} = (4.54/235.4) × 10⁹ Pa   19.3 MPa

     = 19.3 MPa

Deflection = PL³/48E I_{zz}

E =

>> E   = (0.734 × 9.81 × 0.3³)/(48 × 0.0012 × 976.8 × 10⁻¹²)   3,440 MPa

     = 3,440 MPa

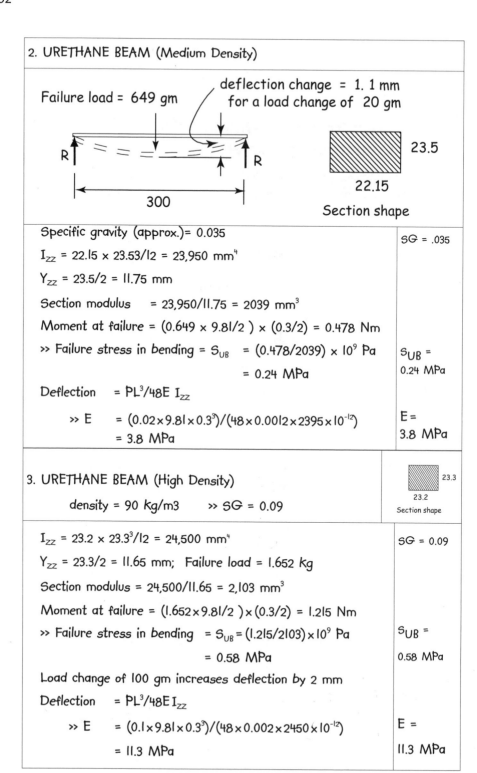

2. URETHANE BEAM (Medium Density)

Failure load = 649 gm

deflection change = 1. 1 mm
for a load change of 20 gm

23.5

22.15

Section shape

300

Specific gravity (approx.)= 0.035

I_{zz} = 22.15 × 23.53/12 = 23,950 mm^4

Y_{zz} = 23.5/2 = 11.75 mm

Section modulus = 23,950/11.75 = 2039 mm^3

Moment at failure = (0.649 × 9.81/2) × (0.3/2) = 0.478 Nm

>> Failure stress in bending = S_{UB} = (0.478/2039) × 10^9 Pa

= 0.24 MPa

Deflection = $PL^3/48E\ I_{zz}$

>> E = (0.02×9.81×0.3^3)/(48×0.0012×2395×10^{-12})

= 3.8 MPa

SG = .035	
S_{UB} = 0.24 MPa	
E = 3.8 MPa	

3. URETHANE BEAM (High Density)

density = 90 kg/m3 >> SG = 0.09

23.3

23.2

Section shape

I_{zz} = 23.2 × 23.3^3/12 = 24,500 mm^4

Y_{zz} = 23.3/2 = 11.65 mm; Failure load = 1.652 kg

Section modulus = 24,500/11.65 = 2,103 mm^3

Moment at failure = (1.652×9.81/2)×(0.3/2) = 1.215 Nm

>> Failure stress in bending = S_{UB} = (1.215/2103)×10^9 Pa

= 0.58 MPa

Load change of 100 gm increases deflection by 2 mm

Deflection = $PL^3/48E\ I_{zz}$

>> E = (0.1×9.81×0.3^3)/(48×0.002×2450×10^{-12})

= 11.3 MPa

SG = 0.09	
S_{UB} = 0.58 MPa	
E = 11.3 MPa	

4. FOAMCORE 300 x 375 sheet weighs 70 gm >> SG = 0.115	5.4 23

$I_{zz} = 23 \times 5.4^3/12 = 302$ mm^4	SG = 0.115
Failure load = 870 gm over 200 mm centres	
$Y_{zz} = 5.4/2 = 2.7$ mm; Section modulus = 302/2.7 = 118 mm^3	
Moment at failure = $(0.870 \times 9.81/2) \times (0.2/2) = 0.427$ Nm	
Failure stress in bending : $S_{UB} = (0.427/118) \times 10^9$ Pa	$S_{UB} =$
$\qquad = 3.6$ MPa	3.6 MPa
Deflection: 317 gm load gives 2.5 mm deflection change	
>> $E = (0.317 \times 9.81 \times 0.23)/(48 \times 0.0025 \times 302 \times 10^{-12})$	E =
$\qquad = 687$ MPa	687 MPa

5. NEWSPRINT (all dimensions are in mm units)

Tensile test specimen	Plain collar crushing test specimen	Pleated collar crushing test specimen

(a) Tensile test	
Failure loads (test specimen shown):	
Machine direction 1.402 kg	
Cross direction 523 gm	
$\quad S_{UMD}$ = failure stress – machine direction	
$\qquad = (1.402 \times 9.81)/(6.8 \times .0625 \times 10^{-6})$ Pa	$S_{UMD} =$
$\qquad = 32$ MPa	32 MPa
$\quad S_{UCD}$ = failure stress – cross direction	
$\qquad = (0.523 \times 9.81)/(6.8 \times .0625 \times 10^{-6})$ Pa	$S_{UCD} =$
$\qquad = 12$ MPa	12 MPa

5. NEWSPRINT (continued)	
(b) Collar Crushing Strength Comparison	
Two paper tubes – 130 mm long	
Smooth tube crushes at 440 gm	
Pleated tube crushes at 616 gm	Crushing
Cylinders were constructed from a sheet of 130 wide x 405 long newsprint	strength
Crushing failure, measured in Load per Metre (LPM) applied at the edge of the tubes	
Smooth tube $LPM_S = (0.44 \times 9.81/0.405 = 11$ N/m	$LPM_S = 11$ N/m
Pleated tube $LPM_P = (0.616 \times 9.81)/.405 = 15$ N/m	$LPM_P = 15$ N/m
The pleated cylinder used one thickness of paper, and the smooth cylinder used $0.405/(\pi \times 0.085) = 1.52$ thicknesses. Clearly pleating is a more efficient use of paper	

(c) Slender Column Test

A well consolidated tube of newsprint was constructed from approximately 1/3 broadsheet

Thickness $t = 18$ layers x 0.0625 = 1.125 mm

$I_{zz} = \pi [D^4 - (D - 2t)^4]/64 = 212$ mm^4

Area $= \pi [D^2 - (D - 2t)^2]/4 = 27.4$ mm^2

Radius of gyration $r = (I_{zz}/Area) = (212/27.4) = 2.8$ mm

Slenderness ratio $= L/r = 495/2.8 = 176.8$

Critical slenderness ratio

$= (2\pi^2 E/S_y)^{0.5} = (2 \times 3.14^2 \times 5.8 \times 10^9/12 \times 10^6) = 97.7$

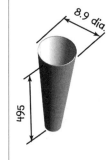

The value of yield strength (S_y) was taken from published data. The value of elastic modulus (E) was estimated from a "slender column" evaluation of this paper tube's performance. Clearly, since the critical slenderness is less than the actual slenderness of the column, the slender column evaluation is valid

Slender column formula

$P_{CR} = (\pi^2 EI_{zz})/L^2 = $ (approximately) 5 kg (measured)

$\gg E = (0.495^2 \times 5 \times 9.81)/(3.14^2 \times 212 \times 10^{-12}) = 5.8$ GPa

E = 5.8 GPa

6. BAMBOO SKEWERS (Kebabs)

Diameter = 2.6 mm; Length = 250 mm; SG = 0.81

(a) Bending (simply supported on 200 mm centres)

185 gm load change resulted in 12 mm change in deflection

$I_{zz} = \pi [D^4]/64 = 2.24$ mm^4

Deflection $= PL^3/48E\ I_{zz}$

$\gg$ E $\quad = (0.185 \times 9.81 \times 0.2^3)/(48 \times 0.012 \times 2.24 \times 10^{-12})$

$\quad = 11.3$ GPa

Note : Bamboo skewer did not break, but bent sufficiently to slip out of supports

(b) Buckling 250 mm length buckled at 400 gm load

Slender column formula $P_{CR} = (\pi^2\ EI_{zz})/L^2 = 400$ gm

$\gg E_{buckling} = (P_{cr} \times L^2)/(\pi^2 \times I_{zz})$

$\quad = (0.4 \times 9.81 \times .25^2)/(3.14^2 \times 2.24 \times 10^{-12})$

$\quad = 11.1$ GPa

Note : Published E for bending of bamboo flooring = 6.5 Gpa

SG = 0.81	
E = 11.3 GPa	
$E_{buckling}$ = 11.1 Gpa	

7. DRINKING STRAWS

Outside diameter D = 5.15 mm; Thickness t = 0.12 mm;
Length L = 165 mm
Failure load (buckling) = 300 gm (direct testing)

$I_{zz} = \pi [D^4]/64 = 6.0$ mm^4

Slender column formula $P_{CR} = (\pi^2\ EI_{zz})/L^2$

$\quad\quad\quad\quad\quad\quad\quad = 0.3 \times 9.81 = 2.94$ N

$\gg$ E $= (0.165^2 \times 2.94)/(3.14^2 \times 6 \times 10^{-12}) = 1.35$ Gpa

Slenderness ratio check used S_y = 35 MPa @ 20°C (source : Ashby, 1992) $\gg$ critical slenderness = 28.4 compared to actual slenderness of 87. Hence slender column assumption is valid

Solid density of straw – 32 straws weighed 12 gm

Section area $= \pi [D^2 - (D - 2t)^2]/4 = 1.9$ mm^2

$V_{solid} = 32 \times 1.9 \times 165 = 10^4$ mm^3

$V_{straw} = \pi \times 5.15^2/4 \times 165 \times 32 = 1.1 \times 10^5$ mm^3

Solid density $= 0.012 \times 10^9/10^4$ kg/m^3 $\quad\gg$ SG = 1.2

Straw density $= 0.012 \times 10^9/1.1 \times 10^5$ kg/m^3 $\quad\gg$ SG = 0.11

E = 1.35 GPa

4.7 Chapter Summary

This chapter is devoted to exploring the mechanical properties of unconventional engineering materials that might be specified for MaT project construction.

- The main types of materials examined and tested are: balsa wood, newsprint, urethane foam, foamcore, spaghetti, drinking straws and bamboo skewers. Wherever appropriate some background information is provided about these materials.

- A wide range of available adhesives is defined and described.

- Designers and participants of MaT projects are encouraged to carry out simple tests to determine mechanical properties of the specific materials specified in projects. Examples are described of simple testing apparatus and test procedures.

- For all the unconventional engineering materials tested tables of data are provided, together with some comparative mechanical properties of conventional engineering materials.

5

MaT PROJECTS FOR STATIC LOAD-BEARING STRUCTURES

Misce stultitiam consilis brevem: Dulce est desipere in loco. [Mix a little foolishness with your prudence: it is good to be silly at the right moment]
Horace, *Odes*, Book 4, 27 BCE

Professor Rosalind Williams in her book[5.1] about the influence of social and technical change on the academic culture at MIT remarks about her maternal grandfather, Warren Kendall Lewis, the founding professor of chemical engineering at that institute:

...Grandpa Lewis thought of himself as a farm boy. As professor at Massachusetts Institute of Technology, he would often try to get some engineering point across by telling a funny story about a country schoolteacher or about an apocryphal Cousin Willie. A teacher's job, he said, was to "get the fodder down on the ground where the calves could get at it."

Although Williams' book is partly about the gradual blurring of boundaries between science and technology at MIT, she admits that designing and building structures and machines is still the domain of the engineer. MaT projects provide excellent opportunities for engineering undergraduates to experience encapsulated engineering design and construction, with the accompanying, almost instant, gratification of testing their ideas in "practice". The educator's dilemma remains that of making the experience one of learning as well as gratification. To this end the following examples present some sample formats for designing and constructing static MaT projects.

5.1 Sample MaT Project Problems and Presentation

The following MaT projects are representative examples for use with small-team design problem solving with engineering undergraduates. P5.1 is a MaT project presented as a formal design specification for a design team. This sample MaT project is intended to illustrate some of the key features of the whole MaT project learning experience. These include:

5.1 Williams, R. (2002).

- An interesting and challenging problem within the skill and experience domain of engineering undergraduates encountering engineering design synthesis for the first time.

- Exposure to the progression from problem to solution in design, that is,

  – problem description,

  – design criteria and design constraints,

  – designer/client interactions and guidance through dialog in planned consultation sessions,

  – deciding on a course of action based on some progressive idea generation procedure,

  – real-life testing of a structure constructed from design ideas,

  – teamwork in preparation and presentation of design ideas, and

  – formal reporting and evaluation of design performances.

- Dealing with success (and occasionally failure) in public exposure of design ideas.

P5.1 *DESIGN PROJECT - INSTRUMENT GUARD*

I. Introduction

This is an introductory Make-and-Test (MaT) project in engineering design. Although one purpose of this project is to introduce some basic ideas of engineering design in the guise of an interesting and challenging problem, the essence is to have some fun while experiencing some design learning.

2. The problem

Imagine that your design team has been entrusted with the protection of a precious and delicate instrument. For the purposes of this project the instrument will be represented by a fresh egg, a reasonable metaphor for a delicate instrument, because, in the first instance, design takes place in the imagination. Your design team has been advised that the instrument is to be protected from falling objects that may damage it while deployed in the field for measuring and collecting data. The client for the instrument guard is an instrument supplier (IS), whose clients require some guarantee of performance for the instrument guard. To gain the confidence of instrument clients the IS has requested a demonstration prototype design, to be field tested in the presence of some prospective clients.

3. The Challenge

A structure is required to protect a fresh egg from a falling mass. The protecting structure, the instrument guard, is to be constructed from newsprint and adhesive tape. To simulate the nature of the falling mass, a single standard wire cut house brick (mass 3360 gm) will be dropped from a height of 300 mm vertically above the highest point of the instrument guard. This height will be measured from the surface on which the instrument (egg) rests.

The egg must be placed approximately centrally in a pastry tray (the instrument stand) of 300 mm diameter with sides of 25 mm height above the base. The instrument guard must not encroach on or touch any part of the pastry tray in which the egg is housed.

4. Organisation and Assessment

Week	Activity	Value %
1	Project is introduced, design teams are constituted and major issues of the project are discussed with client.	
2	First interview with client. Initial ideas and sketches of the instrument guard are presented to the client. Subsidiary or additional problems may arise during the discussions relating to these initial ideas. Guidance may be sought from client on these.	10
3	Sales presentation to a learned audience of your peers. Design teams will prepare a brief (5 min.) presentation of the main ideas for the instrument guard, together with a scaled (or dimensioned) sketch of the most favoured recommendation to the client. The sketches of these recommended structures are submitted to the client for final evaluation. It is expected that these chosen structures will be constructed and submitted for testing by each team. No subsequent variation from these choices will be permitted.	20
4	Client's representatives test completed structures. Teams submitting a structure substantially different from that shown on their submitted sketch will be penalized 30% in assessment.	
5	Design reports are submitted to client. The report must include: idea sketches – idea-logs; a brief history of the development of the chosen design – design diary; any material testing and calculations or analysis involved as the design progressed; a clear statement of the contribution of each team member; a team vote for the most elegant design submitted in concept and construction. (Votes are to be on a scale of 1 to 10 and may include your own team's submission.) Photographs of the structures tested will be displayed on the client's design Web site.	30
6	Feedback from client including evaluation of instrument guard performances (30% maximum) and constructional elegance (10% maximum).	40

This is a team design exercise, with four designers per team. Team members may be chosen freely. The *Activity Table* above sets out the programme and assessment for this project.

5. Additional Specifications and Instructions

The prototype structure must be constructed from newsprint. Two standard sheets of a broadsheet newspaper (each 600 x 810 mm approximately). In addition 2 metres of 20 mm wide adhesive tape (or equivalent) may be used. No other materials of construction are permitted.

The main criterion of performance is the survival of the instrument (egg). In addition, given the vertical height of the instrument guard measured from the surface on which the instrument support rests is H mm, and the mass of the instrument guard has a total mass of M gm, a further criterion of performance is the performance parameter H/M.

The "best" instrument guard will be deemed to be the one that successfully protects the egg, as well as designed and constructed to have the largest value of H/M while doing so. Additional weight in assessment will be awarded for constructional elegance.

The *Instrument Guard* project is only a representative sample of the great variety of MaT projects that may be generated, using the morphological tables in Chapter 3, or by any other combination and permutation of the project examples offered in the list below. Emphasis on the components of a project can become a design variable in MaT project planning. The proportionate weighting, indicated by the values assigned in the *Instrument Guard* project, are illustrative values only and may be changed to reflect the individual learning preferences of engineering educators planning MaT projects.

P5.2 Construct a structure for supporting a fresh egg. The materials of construction are bamboo skewers (approximately 2.6 mm diameter x 250 mm long) and any two-part room temperature curing adhesive.

Criteria of Performance are:

- Maximum height H of the structure (measured from base to the top of the supported egg. It is desirable to make this measurement as large as possible;

- Total mass M of the structure, not including the egg. It is desirable to make this mass as small as possible, while maintaining the integrity of the structure;

- It is expected that the "best" design will be a structure with the largest value of the design performance index H/M;

- Structural stability is to be tested by tapping the top of the structure, while supporting the egg, with the end of a 300 mm wooden school ruler, held near the top between the thumb and forefinger and released at 45° from the horizontal. This test is to be applied in two directions perpendicular to each other (refer to Figure 3.3).

P5.3 Construct a structure similar to that described in P5.2 from: (a) drinking straws (approximately 5 mm diameter x 1.125 mm thickness), and from (b) dry spaghetti (approximately 1.8 mm diameter), using a suitable two part room temperature curing adhesive. Performance criteria for both of these structures are to be the same as for the structure described in P5.2. Compare the overall performance of these two structures. Can it be predicted which structure will perform better in terms of height-to-mass ratio before they are constructed and tested?

P5.4 Construct a simple structure capable of carrying only its own weight while spanning a chasm (say the space between two desks). The material of construction is a piece of 300 mm x 300 mm x 5.4 mm foam-core. No other materials may be used. The criterion of performance is the span of the structure. It is desirable to make this span as large as possible. It is useful to consider the various alternatives to this problem before commencing construction. Is it possible to predict the maximum length possible (i.e. carry out a simple thought experiment)?

P5.5 Construct a structure similar to that of P5.4, using three pieces of balsa, each 6 mm x 100 mm section x 900 mm length. No other materials may be used. The criterion of performance is the same as for P5.4.

P5.6 Imagine a competition in which several design teams are required to design and construct the tallest structure from a single broadsheet of newsprint (600 mm x 810 mm), using only 1 metre of 12 mm wide adhesive tape as jointing material. The structure is to carry a full can of beverage (Coca Cola or similar, 378 ml) and it is to be tested for stability in a manner similar to that described in P5.2. All competing structures will be tested and in case of a tie for tallest structure between competing structures, these will be tested to destruction using increasing numbers of full cans. The tallest structure capable of carrying the largest number of cans stably will be deemed the winner.

P5.7 Short paper columns usually fail in local crushing. The variables of construction are shown in Figure P5.7. Carry out the optimisation of the paper tube to establish the best geometry for carrying a load (*i.e.*, maximise load/unit length of paper used in the structure by varying

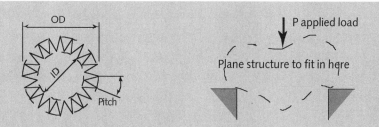

Figure P5.7 Section through paper column   Figure P5.8 Space to fit structure

the geometry of the pleating for a given length of paper). The pitch of the pleating is measured in number of pleats per circumference length (or degrees). Care is needed in construction and adhesive tape may be used for joining the first pleat to the last to form the tube. Height may be maintained at some nominal value (say 100 mm). Optimisation should involve varying the two diameters (*ID* and *OD*) and pleat pitch.

P5.8 A simple planar triangulated structure is needed to support a centrally located load P above simple supports. You may use as many *struts* (compressive members) and *ties* (tensile members) as you wish. However, the objective is to find the geometry of the minimum weight structure. It may be assumed that the ties take 100 times as much load as the struts (due to *buckling*[5.3]; see Figure P5.8).

P5.9 Design and construct a structure to carry the weight of as many copies of the local telephone directory as possible. One half sheet of broadsheet newspaper may be used as well as 0.5 metre of 12 mm wide adhesive tape. Minimum height is 100 mm.

P5.10 A special cargo (a fresh egg) is to be transported from a height of 15 m to the ground. Design and construct a transport vehicle to ensure safe transport (without damage) of the cargo. One sheet of a broadsheet newspaper may be used to construct the transport vehicle. One meter of 12 mm wide adhesive tape may be used. Also, up to 10 small paper clips, and 0.5 metre of Tex 50 (or similar)[5.2] cotton thread may be used. No other materials of construction are permitted. Success of safe delivery of cargo is mandatory. Other criteria of performance are:

• Simplest construction used (least number of components, simplicity of interconnectedness),

• Lightest vehicle (not including cargo).

5.2 Refer to Chapter 4, Table 4.6 for approximate breaking strenth of cotton thread.
5.3 For a description of column failure behaviour refer to Appendix A1, *A Primer on Mechanics*.

P5.11 Design and construct a structure capable of spanning a 2 m wide chasm (space between two simple supports). Materials of construction consist of one sheet of 25 mm thick 90 kg/m³ polyurethane foam, and a two-part room temperature curing adhesive. The structure is required to carry one full can of some beverage (Coca Cola or similar, 378 ml). Criterion of performance is weight of structure (to be minimum). Referring to the performance characteristics of polyurethane foam in Chapter 4 (Section 4.3 and Table 4.6), estimate the mass of the lightest structure that will satisfy the design requirement. of this project.

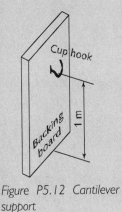

P5.12 Design and construct a simple structure to carry the weight of one full can of some beverage (Coca Cola or similar, 378 ml). A particle-board sheet 1 m wide x 1.2 m height will be provided with a cup hook located centrally at 1 m from the base of the board. The required structure may be attached to this hook and may rest against any other part of the board. No other attachments may be used. One sheet of broadsheet newsprint may be used as well as 1 m of 12 mm adhesive tape. The lightest structure will be deemed the best.

Figure P5.12 Cantilever support

5.2 Case Examples of Static MaT Projects

5.2.1 Balsa Bridge

I. The Problem

Given two sheets of balsa wood 100 mm wide, 1200 mm long, 4 mm thick, design and construct a bridge to carry only its own weight. No other materials of construction are permitted. The sag of the bridge (displacement measured in the gravity direction from the horizontal line joining the simple supports) must not exceed 1/5 span. The best design will have the longest span.

2. The Solution

In MaT projects, student designers are required to submit sketches and descriptions of ideas at regular intervals while their planning, design and construction of their project model proceeds. They are also advised that these idea-logs of the development of their designs should be submitted, as far as possible, in their unsanitised "raw" form. For the balsa bridge the following collection of sketches represents a collage of student idea logs (Figure 5.1).

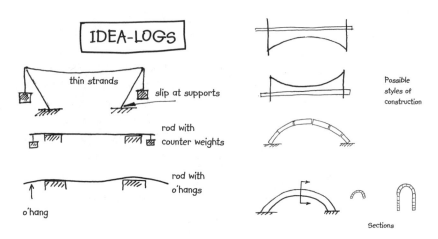

Figure 5.1 Idea-logs generated by students during initial design planning

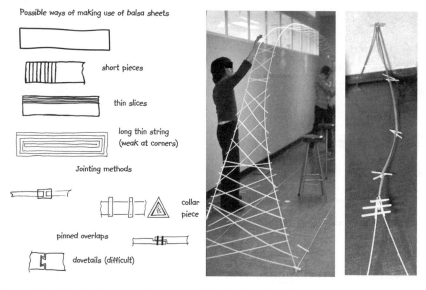

Figure 5.1 Idea-logs(continued): cutting and jointing ideas

Figure 5.2 A complicated bridge being constructed and a simpler "string" bridge

The photographs of Figures 5.2 through 5.6 show the wide range of bridge constructions used by students. Most submissions ranged from approximately 3 metres to about 15 metres in length, with an overall average of a 7 metre span. The "best" submission was shared between two teams, both with "thin string" type bridges. These two submissions used a specially designed "shaving" device to cut very thin slices of balsa for their "strings". Friction joints, shown in Figure 5.7, joined the strings.

Figure 5.3 Simple suspension model

Figure 5.4 Simple arch model

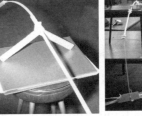

Figure 5.5 A triangulated suspension bridge

Figure 5.6 Overhung supports

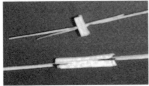

Figure 5.7 "Thin string" joints

Figure 5.8 More complex joint models

3. Some Lessons Learnt from this Project

It is instructive to ask engineering colleagues to "guess" the longest span for a bridge constructed according to the rules of this specific MaT project. Often it will be difficult to get a response without some deference to authority (data on balsa properties), or formal analysis. In my experience with colleagues who have not tackled this problem previously, and who are prepared to guess, the answers range from 3 metres to 20 metres. None have yet guessed 28 metres (the length of the string bridges when these teams ran out of construction space).

Perhaps the most significant experience learnt from this exercise is that some experimentation and some wide-ranging idea generation go a long way towards the design of simple and elegant solutions. Almost without exception, the teams that produced designs with longer than average spans, also produced extensive idea sketches and recorded substantial experimentation with their materials. Another important issue in this design was the use of careful manufacturing techniques in joint construction.

5.2.2 Balsa Structure

I. The Problem

The design and construction of a planar structure, capable of supporting a minimum load of 50 N. Due to space constraints the completed structure is required to fit into the space whose boundaries are indicated in Figure 5.9. The completed structures will be tested to destruction. The criterion of performance will be the ratio of failure load W to structure mass M. It is desirable to make the performance index W/M as large as possible.

Unlimited quantities of balsa may be used in constructing this structure. As well, a suitable balsa glue, or, alternatively, a two-part room temperature curing epoxy glue may be used in the construction.

2. The Solution

Figure 5.10 shows a photo of a sample of structures presented as solutions to this project. Student designers were asked to carefully establish the masses of their structures and these masses were used in evaluating the performance indices. Figure 5.11 shows selected structures with their mass, failure load and some indication of their respective failure modes.

3. Some Notes on Performance and Lessons Learnt

These structures represent the inspired (nonanalytical) thinking of young engineering undergraduates. In many cases the students knew about simple beam behaviour, but had no formal training in the behaviour of framelike structures, columns or arches. Many of the submitted structures resulted from long hours of trial and error.

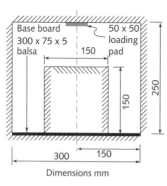

Figure 5.9 Loading diagram indicating exclusion zones into which the structure is forbidden to intrude

Figure 5.10 A sample of submitted structures

(a) Structure mass 84.8 gm;
failed at 530 N

(b) Structure mass 171.4 gm;
failed at 653 N

(c) Structure mass 87.7 gm
failed at 522 N

(d) Structure mass 48 gm
failed at 492 N

(e) Mass 28.5 gm,   (f) Mass 28.5 gm,   (g) Structure mass 46.3 gm
failed at 505 N   failed at 592 N   failed at 134 N

Figure 5.11 Representative sample of submitted structures with their respective masses and failure loads

- *Figure 5.11(a)* – A heavy structure in which the topmost beams are overdesigned and failure in the columns, that form the legs of the structure, determines overall load-bearing capacity;

- *Figure 5.11(b)* – A very heavy triangulated structure, in which the lower part of the structure is much more robust than the top section, where eventual failure occurred;

- *Figure 5.11(c)* – Another simple, heavy structure, where the failure commenced in delamination of poorly glued components;

- *Figure 5.11(d)* – This a more elegant triangulated archlike structure, where the failure occurred at weak or poorly constructed joints;

- *Figure 5.12 (e)* – The failure mode of this elegant, complex, lattice structure is shown in Figure 3.8, where the left leg and right beam support fail together. Clearly, this is a more efficient structure than the ones shown in the earlier figures;

  Figure 5.12 (f) –This simple and elegant arch structure was the most efficient structure submitted. Apart from its constructional simplicity and efficient use of materials it had fewer joints exposed to failure than its competitors. It is conjectured that this team of designers had not only inspiration, but also expert advice in completing their design;

- *Figure 5.13* – As a contrast to the elegant arch of Figure 5.12, this simple structure made very poor use of materials and consequently had the poorest load-to-mass ratio of 2.9.

Table 5.1 gives relative performance indices for these structures. The essential lessons learnt from this exercise were the need for care and attention to planning as well as detail design. Most failures seemed to occur at poorly constructed joints or poor appreciation of structural behaviour (*i.e.,* insufficient experimentation).

Table 5.1 Performance indices for balsa structures			
Figure	Mass of structure M gm	Load at failure W N	W/M
5.11(a)	84.8	530	6.3
5.11(b)	171.4	653	3.8
5.11(c)	87.7	522	6.0
5.11(d)	48.0	482	10
5.11(e)	28.5	505	17.7
5.11(f)	28.5	592	20.8
5.11(g)	46.3	134	2.9

5.2.3 Paper Column

I. The Problem

This project has an almost fairy-tale-like beginning. Once upon a time, in bygone (heady?) days of university life, engineering undergraduates had access to *drawing-boards* and *Tee-squares* and *slide-rules*. The phrase *"back to the drawing-board"* has more or less outlived the now almost obsolete drawing-board. Still, even though the drawing-board in engineering schools may be a thing of the past, there are other opportunities for supporting large flat load-spreaders on paper columns. A stiff timber board that does not bend under modest loads could be utilised for such a project. The essence of spreading the load, as evenly as possible, over the several columns to be designed, is a key element of paper column MaT projects. If the project is to emulate true column behaviour as closely as possible, then eccentric or lateral loads should be avoided. Such loads are generally imposed when the supported load deforms at the supports. Large telephone directories and particle boards tend to behave in this unhappy way. These early caveats set the scene for planners of paper column MaT projects. Because the designer's life is beset with sufficient uncertainties even in simple design problems, well-planned MaT projects should try to avoid unnecessary complications.

This project was set in the days predating computer workstations and *Autocad*. The task was to design and construct a set of paper columns to stably support some numbers of drawing-boards. The standard A2 size boards made of red pine had a mass of approximately 4 kg each. Only one single sheet of a broadsheet newspaper and a metre of 12 mm wide adhesive tape were the permitted materials of construction.

Performance criteria for this project were:

- Height H of the paper columns (more than one column would be needed to stably support the stack of drawing boards) above a flat surface. It was desirable to make H as large as possible;

- N the number of drawing-boards in the stack supported stably by the paper columns. Stability is to be tested by a "light tap" on the edge of the drawing-board stack by a Tee-square. It is desirable to make N as large as possible;

- S the number of supports of equal length used. It is desirable to make this number as small as possible.

A composite performance parameter for this project was $H \times N/S$.

2. The Solution

Figure 5.12 shows a small, representative sample of typical submissions for this MaT project.

- *Figure 5.12(a)* – A multi tube design, with the advantage of being more compact and permitting better distribution of load per unit circumference of the column. Because the several tubes share the load, there is less influence on performance from manufacturing inaccuracies in individual columns;

- *Figures 5.12(b)* – Columns with increased end diameters to spread the loading there. This was a popular approach taken by designers. In the case shown, skillful manufacturing ensured that the

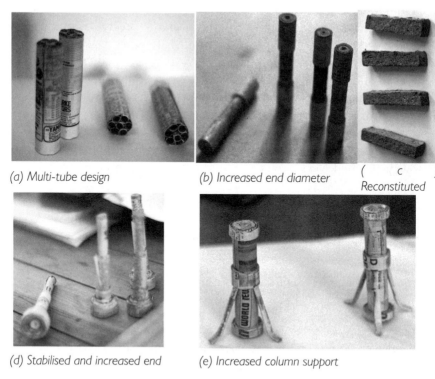

(a) Multi-tube design   (b) Increased end diameter   (c)
Reconstituted

(d) Stabilised and increased end   (e) Increased column support

Figure 5.12 Representative sample set of entries to the paper column project

increased size ends were sufficiently compact to provide genuine
load spreading;

- *Figure 5.12(c)* – A radical design where the sheet of newsprint was
crushed to pulp and reconstituted as solid papier maché columns;

- *Figure 5.12(d)* – Another entry with increased column base. In this
case the wide base was also intended to provide some lateral stability. Sadly, the poor compaction of column ends significantly
reduced their effectiveness either as load spreader or stabilizer;

- *Figure 5.12(e)* – This design attempts to provide some mid-column support to (it is hoped) reduce the chances of midcolumn
buckling. Unfortunately, the supports at mid-column are far too
flimsy for the overall column thickness, and they will have little
or no influence on failure load;

- *Figure 3.1* – A simple but effective pleated column design submission to this project is shown in Chapter 3 (not shown here).

Figure 5.13 Paper columns under test with increasing numbers of drawing boards. This entry managed to support 27 boards (almost) just prior to collapse

Figure 5.13 shows photographs of columns under test as the drawing-board tower progressed to failing these columns. An added engaging feature of this project was the suddenness of column collapse, accompanied by a crashing down of the carefully assembled column of drawing boards.

As Hilaire Belloc noted in a cautionary verse *About John*:[5.4]

John Vavassour de Quentin Jones
Was very fond of throwing stones,
At horses, people, passing trains,
But especially at window panes.

Like many of the upper class,
He liked the sound of broken glass…

Most engineers (it is hoped) have few opportunities to witness the static failure of their designs. Hence, the inclusion of the safe, but noisy and/or spectacular, destruction of static MaT project models can provide a helpful degree of engagement in these projects. Table 5.2 gives some comparative performances for the designs shown in Figure 5.12.

3. A Cautionary Note on Performance

Due to this project's apparent simplicity, designers perceived that the influence of constructional detail would not significantly influence performances. Paper columns could be constructed relatively easily with a minimum of constructional skill. However, as the sample entries shown in Figure 5.12

5.4 Belloc (2002).

Paper column design	Height (H) mm	Boards supported at failure (N)	Number of support-ing columns (S)	H x N/S
5.12(a)	150	12	6	300
5.12(b)	100	32	4	800
5.12(c)	100	400	4	10,000
5.15(d)	120	16	4	480
5.15(e)	150	27	4	810
3.1	120	6	3	240

Table 5.2
Comparative performance of a selection of paper columns submitted

indicate, design performance did indeed become a function of construction-al skill. Care in the construction and compaction of columns, as well as the care in constructing thickened column ends, significantly influenced per-formances. The "reconstituted" papier-maché entry shown in Figure 5.12(c) was clearly outside the intended spirit of the project. However, it was, equally clearly, a case of "thinking outside the square". In general, cre-ative freedom should not be stifled in these projects by undue commitment to some design agenda. In the case described this entry was simply treated as belonging to an alternate class of designs. It became necessary to test this design in a mechanical testing machine.

5.2.4 Urethane Foam Structure

1. The Problem

MaT projects using structural materials unfamiliar to most engineers, and especially to undergraduate designers, tend to bring a sense of challenge and excitement to the design experience. In this project sheets of medium den-sity polyurethane foam, 600 mm long, 300 mm wide and 25 mm thick, was provided to each design team. Sufficient material was provided to permit some structural experimentation with the material. Design teams were asked to design and construct a load-bearing structure that met the space constraints imposed by the loading diagram shown in Figure 5.14. Designers were advised that the structure would be loaded to failure and the best structure would be the one with highest failure load-to-mass ratio.

Criteria of performance:

- *Failure load* – P is the load in Newton at failure. It is desirable that the structure should achieve as high a value of P as possible.

- *Mass of structure* – M is the mass of the structure in gm. It is desir-able to make the mass M of the structure as low as possible.

- *Composite parameter* – P/M should be as large as possible.

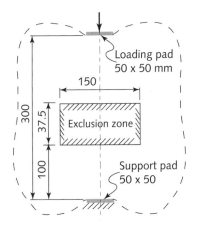

Figure 5.14 Loading arrangement for foam structure. Dimensions are in mm *Figure 5.15 A sample structure submitted*

2. The Solution

Figure 5.15 shows a sample of structures submitted by students. It was expected that student teams, participating in this MaT project, would use about half of the sheet of urethane foam to carry out testing and experimentation with this unfamiliar engineering material.

As may be observed from the submitted structures, some students did indeed do some testing before "carving out" (perhaps the most appropriate phrase to describe the general urethane foam manufacturing process) their submissions. Some students chose to cut their sheet of foam in half and to submit two different models.

It is illuminating to examine the few samples chosen from the many structures submitted. A feature of the mechanical properties of the foam is that it has better compressive strength than tensile strength. As a consequence, beams are very sensitive to local tensile stress concentrations.

- *Figure 5.15 and Figures 5.16(a) through 5.16(e)* – Teams submitting these structures were clearly aware of the need to support the applied loads by some triangulated structure and, as far as possible, to make best use of the properties of the foam. The arrow in Figure 5.16(b) points to the location of commencement of failure at a local stress concentration caused by manufacturing damage. A team with some degree of artistic flare submitted Figure 5.16(c), and Figures 5.16(d) and 5.16(e) are slightly aberrant submissions by teams who thought along similar lines. The aberrant feature of Figures 5.16 (d) and 5.16(e) is the material protruding above the loading pad in both structures. This excess material cannot carry any load and hence it makes these structures less effi-

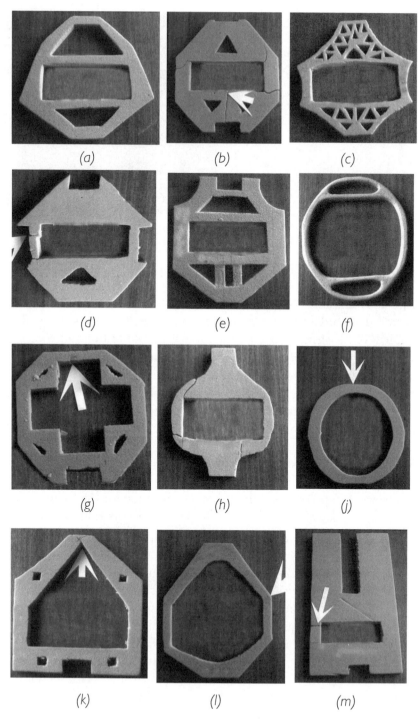

Figure 5.16 A representative sample of foam structure entries

cient than would be the case if these bits of material were to be removed.

- *Figure 5.16(f)* – Presumably this structure is designed on the theory that less is better than more. The two oval structures at the top and bottom will play some minimal part in distributing the load, but the majority of the load will be carried in bending by the ring-like side elements of the structure. In defense of this submission it is clear that avoiding stress concentration was a major consideration in its design.

- *Figures 5.16(g) through 5.14(l)* – These are all essentially "ring" structures, and they indicate some weakness in understanding of the way loads are distributed in such structures. All of these failed in bending tension at local stress concentrations.

- *Figure 5.16(m)* – This submission deserves a special mention as an aberrant structure. This team decided that the whole sheet of foam must be utilized and *hang the consequences*. Clearly, about half the structure could be cut away without loss of load carrying capacity. It may be conjectured that this is yet another case of delimitation of specifications experienced with the paper and balsa bridges described earlier.

5.3 Supplementary Opportunities for Static MaT Projects

The following list identifies some general types of static MaT project opportunities that may be developed by *caring* MaT project administrators.

- *Beams* – Beams are structural elements that support transverse loads. These loads may be applied as "steady" or "dead" loads, or as "varying" or "live" loads. Typically, bridges are designed to carry their own weight (the dead load) as well as traffic (the live load). In MaT projects the easiest form of live load is an impact load, such as experienced in the instrument guard project, described earlier in this chapter. Alternative types of live load may be invoked such as a carriage that transports a payload from one end of a "bridge" to the other. In a sense, this type of MaT project is a more challenging one to develop than those dealing only with dead or impact loads. One challenge is the provision of sufficient mentoring in exploring elementary beam behaviour.[5.5]

5.5 A simple primer on beams is provided in Appendix I.

Further exploration of beam behaviour is offered by specifying challenging support arrangements. The most common forms of support specified are *simple supports*, where the supports may not impose any moments or shear loads on the beam. As an alternative the beam could be specified as a cantilever, where the single support needs to be designed with some care. P5.12 shows an example of such a MaT project. This project requires careful mentoring in the behaviour of cantilevers. Typically the cup hook permits the attachment of a tension member to the backing board and a simple triangulated structure will be one possible solution. The challenge then is to design the structure to carry the load at as large a distance from the board as possible.

The design requires understanding of both tensile and compressive (column) behaviour of structural elements. Further MaT project opportunities are offered by the specification of unconventional materials such as those described in Chapter 4.

- *Multi-dimensionally loaded structures* – These include three-dimensional frames with an opportunity for exploring *structural distillation*.[5.6]

- *Twisting beams* – These are beamlike structures with a superimposed twist or torsion applied. The balsa model example shown in Figure 3.6 is a typical twisting beam.

- *Beam columns* – These are beamlike structures, where both axial and transverse loads need to be considered,[5.7]

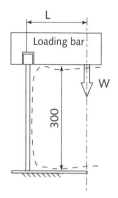

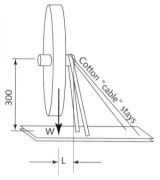

Figure 5.17 A beam column with a chain dotted exclusion zone

Figure 5.18 The London Eye – cable stayed during construction (Photo by author)

Figure 5.19 Schematic representation of the "Eye" during installation

5.6 Refer to Samuel and Weir (1999), Section 1.3.
5.7 See also Gere and Timoshenko (2002).

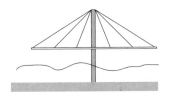

Figure 5.20 Cable-stayed bridge platform (schematic)

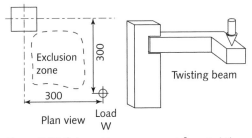

Figure 5.21 Schematic arrangement for a twisting beam, showing a notional exclusion zone

- *Cable-stayed structures* – In these structures a significant part of the load is carried by cables. Figure 5.19 shows a schematic cable-stayed bridge section. The advantage of these structures is that they are usually lighter than equivalent suspension bridges. Moreover, their assembly during construction is simpler than that of suspension bridges.

P5.13 Figure 5.17 shows a schematic stylised view of a type of crane. A model structure is to be designed and constructed from balsa wood. It is desirable to make the distance L and the load W as large as possible. There is no restriction on the quantity of balsa wood or the type of glue used in constructing the model. However, the composite performance index for this exercise is $W \times L/M$, where M is the mass of the structure including loading bar at the top and the base support. The form of attachment of column head to the loading bar is not restricted to that indicated in the figure.

P5.14 Figure 5.18 is a photo of the *London Eye*, the British Airways-owned ferris wheel constructed for the Millennium Expo. During construction the wheel is lifted from the horizontal position to its final vertical position by cables. Once in its final position the wheel rests on the drive system. However, a model is needed to explore the character of the lifting structure. Figure 5.19 shows a type of simple model to be designed and built from balsa wood. Cotton thread (Tex 90 or similar), attached to the top of the angled *A-frame* support, may be used to stabilise the structure. The performance index for this structure is the same as for P5.13, however, in this case the distance L is measured from the top of the A-frame to the load W.

P5.15 Figure 5.21 shows a "twisting beam" in its support with an exclusion zone into which the structure may not intrude. In this case it is desirable to make the load W as large as possible and the performance index is W/M, where M is the mass of the beam structure. A variety of materials may be explored using this type of twisting beam exercise, including balsa wood, urethane foam and foamcore. There should be no restriction on the quantity of materials or the type of glue used in construction.

5.4 Chapter Summary and Caveats

In this chapter some case examples of static load-bearing MaT projects were reviewed. Balsa wood, paper and urethane foam were explored as case studies of the use of some unconventional engineering materials. Two special caveats are offered below for the disciplined design of challenging and engaging MaT projects

The learning experience of handling, testing and designing with unconventional engineering materials can be substantially enhanced by care in mentoring student teams. In this case the mentoring could have included some basic ideas on frame behaviour. This is a common feature of MaT projects, inasmuch as they are essential learning experiences for students and academic staff alike.

In general, it is possible to design MaT projects in which student designers are expected to explore engineering issues of some specific individual interest. The flying machine MaT project (*DYDaF* – 1973) described in the case studies in Chapter 7 is one such example. Almost all design academics have been subjected to offers of "good" or occasionally "great" design projects from colleagues. Some of these "offers" are of problems that the "helpful" colleagues may have explored for themselves and where they have developed "design solutions" that satisfied their specific need, often to the exclusion of alternative design possibilities. I too have been exposed to such offers during my academic life at Melbourne. In some cases the colleague simply wanted the student designers to come up with the same solution that they themselves had developed. These types of design projects tend to generate a kind of *game against specifications*, rather than the more realistic *game against nature* which is the real challenge of design problems. In really worthy MaT projects the student (and academic) wit should be pitted in a problem-solving process that is a *game against nature*.

The use of undisciplined MaT projects incurs the inherent danger of exceeding student capabilities to such an extent that instead of becoming engaged with the problem, they become disengaged from engineering design.

6
DYNAMIC MaT PROJECTS: THINGS THAT GO "BUMP"

Eppur si mouve [and still it moves]
Galileo Galilei, 1632;

From ghoulies, gosties and long-legetty beasties
And things that go bump in the night,
Good Lord, deliver us!
The Cornish or West Country Litany
Francis T. Nettleinghame, 1926

Galileo Galilei (1564–1642) was probably the greatest scientist of the current era. The Encyclopedia Britannica lists, among his other achievements:

> ...*His insistence that the book of nature was written in the language of mathematics changed natural philosophy from a verbal qualitative account to a mathematical one in which experimentation became a recognised method for discovering the facts of nature.*

In Galileo's time natural philosophers held the view (belief based on the authority of Aristotle only) that objects moved at speeds proportional to their masses, a view attributed to Aristotle's philosophical account of nature. In an early experiment, reputedly conducted from the balcony of the leaning tower of Pisa, Galileo dropped a musket ball and cannon ball, many times the mass of the musket ball, to see how fast they would fall. He found that both masses reached the ground at about the same time. From this experiment Galileo concluded, within some slight experimental error attributed to air resistance, that this result clearly refuted the inappropriate Aristotelian view of nature about falling masses.

With this, as well as many other similar basic experiments, Galileo firmly established the need and, indeed, the method for experimental verification of intuitive hypotheses about the nature of the universe.[6.1]

Engineering science is almost entirely based on experimental adjustments to mathematical models of observable engineering phenomena. Dynamic MaT projects permit first-hand experimentation with the most basic of engineering phenomena, namely the exchange of energy to perform useful work.

6.1 See also Atkins (2003).

6.1 Energy Sources and Their Measurement

Dynamic MaT projects require energy in one form or another. These were listed in Section 3.2 and are discussed in the same order here.

Thermal Energy

The most common forms of thermal energy sources may be classified as primary(or direct) thermal energy sources and secondary (or indirect) thermal energy sources. Primary thermal energy is delivered by the direct chemical reaction of burning, or oxidation of materials. Typical examples include burning of fuel and, more generally, exothermic chemical reactions. Chemical explosion, nuclear fission and nuclear fusion (the burning of hydrogen to helium), are other forms of primary thermal energy sources. In chemical explosions the rates of pressure rise are so rapid that the energy transfer is very difficult to contain in a mechanically productive way. In artillery, the rapid pressure rise of explosion is used to transfer potential energy of gas pressure to the kinetic energy of the projectile.

Nuclear fission requires very special types of technology to maintain the process under control. The radiation generated from fission products is a special danger associated with this type of primary thermal energy source. Nuclear fusion burns gaseous isotopes of hydrogen (deuterium and tritium) to a plasma (ionised gas) of helium at temperatures nearly ten times the core temperature of the sun (about a hundred million °C). The major difficulty with this type of energy source stems from the lack of appropriate technology and materials to contain the reaction at this high temperature.[6.2]

Secondary thermal energy sources are provided by devices or substances preheated by a primary energy source. Some examples are: electric heating coils, light globes, hot water bottles, bed warming pans (usually heated by hot coals), and steam irons used for pressing cloth. The simplest way of evaluating the output of any of these sources is by direct experimentation.

As an example of evaluating the thermal energy available from a primary energy source, consider the energy provided by a burning candle. The candle burns wax, that has a main constituent of paraffin oil, with an energy content of (approximately) 27 MJ/m^3. The rate of burning is easy to measure experimentally. For example, let us suppose that a standard table candle 37 mm diameter x 18.5 mm long weighing 14 gm takes 5 hr to burn completely. The energy content is 1.99×10^{-5} (volume) x 27 = 539 J. The rate of delivery is 539/(5 x 3600) J/s = 0.3 watt.

A secondary energy source is provided by a 100 watt incandescent electric light bulb. The majority of the electrical energy input to light globe is used

6.2 For further reading see for example Cooper and Kurowski (1996); Hewitt (2000); Nolan (1996); Nuclear Energy Agency (2003) and Xanthanite (1999).

Table 6.1 Energy comparisons for fuels[6.3]	
Fuel	Energy content
Paraffin oil	27.1 MJ/m³
Natural gas	3,730 MJ/m³
Propane	17.6 MJ/m³
Electric resistance heating	3.6 MJ/kWhr

to generate heat, the light energy component being a very small part of the total energy output. Light globes are very inefficient users of energy for converting electrical energy into light energy. The energy input to the light globe filament also comes from an indirect energy source. It has been derived by converting a primary energy supply into electrical energy. The consequence of these various stages of energy transformation is the consistent loss of efficiency in the transformation process. Clearly, the major design challenge involved in employing thermal energy is the efficient capture and transformation of it into some useful work. The 100 W globe will provide an approximate heat output of 0.1 kWhr = 0.36 MJ. Capturing this heat energy and converting it to useful work effectively is a challenging design problem. Table 6.1 gives some comparative energy values for other fuels.

In a manner similar to that for candle burning one might consider the use of hot water (preheated) as a source of energy. Although this is a relatively easily derived source of energy, the recovery of work from such a source is very inefficient.

E6.1 If we could recover, with minimal loss, all the thermal energy contained in a cup of hot coffee and use this energy to raise the water contents in the cup, how high could we raise this mass?

Thermal energy provides heating by radiation, convection or conduction. It may be used to do mechanical work only indirectly by the generation of potential energy in compressed gas, steam or material (metal rod, bimetallic strip[6.4] or shape-memory alloy,[6.5] for example). Heat may be used also to generate kinetic energy in the form of convective currents in a fluid. A warm fluid has lower density than a cold fluid and buoyancy drives the warm fluid to rise above its colder counterpart. In this way a convective current may be appropriately contained to generate kinetic energy of motion within the fluid.

E6.2 Consider two parallel strips of metal, copper and steel, connected as indicated in the diagram of Figure E6.2. If we heat the combination to 150°C above ambient temperature (say 20°C), how much will the tip

6.3 Source: American Petroleum Institute.
6.4 Two parallel strips of metals of dissimilar thermal expansion coefficient (copper and steel for example) connected together.
6.5 See, for example, Otsuka and Wayman (1998); Srinivasan and McFarland (2000).

of the strip deflect and in which direction? How much load could such a deflected strip require to return to approximately its original position while still being maintained at 150°C above ambient temperature? The respective coefficients of thermal expansion are: α (x 10^{-6}/°C): copper = 17, steel = 13.

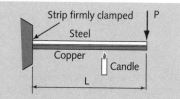

Figure E 6.2 Bimetallic strip. Each material strip is rectangular in section with W width and T thickness

Potential Energy (Gravity Driven)

Table 6.2 Potential energy sources[6.6]		
Energy source	Dimensions (mm)	Mass (gm)
Tennis ball (tournament approved)	66 dia.	56
Golf ball (Royal and Ancient St Andrews approved)	1.68 inch dia. 42.67 dia. 430-515 dimples	45 – 46
Glass marble -standard	15.85 dia.	5.2 (average of 10)
Glass marble – large	25.3 dia.	21
Beverage can (375 ml)	66 dia. x 130 (overall) long	390
Standard three-hole wire-cut clay house brick	225 x 105 x 74 35 dia. holes through 225 x 105 face	3360
Duracell 1.5 V AA battery	14 dia. x 50 length overall	24.4 (average of 10)
Lead shot	Diameter (mm) (average of 50)	gm/100
#2	3.81	32.6
#4	3.3	21
#5	3.05	16.7
#6	2.79	12.6
#7.5	2.41	8.1

Note: Shot is available from most hunting equipment suppliers in both lead and steel . Steel shot is lighter than lead and the relative mass ratio of steel shot compared to lead shot is 1:1.44.

6.6 Source: Direct measurement by author.

The list of available potential energy sources is virtually limitless. Balls, marbles readily available weights (*e.g.,* lead shot), cans of beverage and coins of various denominations represent a small sample of the available items. Coins vary significantly with age and country of origin, consequently they may be specified in terms of their weight and size directly. Metal washers, bolts and nuts are also well suited to be specified as potential energy sources. Table 6.2 provides a brief list of some international standard potential energy sources.

Kinetic Energy

Kinetic energy is the energy of motion expressed as

$$KE = mV^2/2 \,,$$

where m is the mass of the moving body and V is its velocity. Because moving bodies require some form of energy exchange to begin motion we would begin the process of energy exchange with some form of stored or potential energy. Typical examples are :

- A spring (tension, compression or torsion) released from a deflected state to generate motion of a body;

- Pendulum mass used to strike a body to generate motion through impact;

- Compressed gas or liquid released through a jet to generate motion;

- A mass released from some height to generate motion through impact;

- A projectile fired by gas release (air rifle, soda syphon) or by explosion (chemical energy) or by electromagnetic forces (linear motor or "rail gun");

- Strain energy stored in an elastic, released like a catapult to generate the motion of a projectile.

The mighty "California Screamin" roller coaster ride at Disney Themepark in Anaheim, California has the following technical data:

- Acceleration : 0 to 25 m/sec in 4 seconds;

- First hill (highest rise) = 33 m;

- Load is six carriages with a total of 24 passengers.

Two photographs of this ride are shown in Figure 6.1. Starting energy (kinetic) at the bottom of the first hill is $mV^2/2$. After rising to the top of the

Figure 6.1 Photos of "California Screamin" at launch and inversion

first hill the carriage and its load gain some potential energy of *mgh*, where *h* is the height of the first hill and *g* is the acceleration due to gravity. If we assume there are no other losses, then we can equate energies to find the starting velocity at the bottom of the first hill to be

$$V_{launch} = \sqrt{2gh} = \sqrt{2 \times 9.81 \times 33} = 25.5 \text{ m/s}.$$

This implies that for zero forward velocity at the top of the first hill (in the absence of all other forms of losses such as friction and air resistance) the forward velocity of the carriage at the bottom of the first hill must be this value (25.5 m/s). In fact, the actual launch velocity is significantly greater than this value.

E6.3 Making suitable estimates of body mass and carriage mass, calculate approximately how much energy (in joules) the linear induction motors must impart to permit the fully loaded train of carriages to get up the first hill.

E6.4 A standard golf ball is released from a height of 2 m above a carriage, designed to carry only its own mass and the mass of the golf ball, along a level, smooth, hard surface. Estimate the "best possible" starting velocity of the loaded carriage, including the various losses, that might occur during the energy conversion. Some incidental losses might be due to:

• Air resistance on the falling ball;

• Impact losses due to the *coefficient of restitution* being less than unity;[6.7]

6.7 See Appendix A1, section 2.2.

• Rolling resistance on the floor; note that if the wheels of the vehicle are left completely smooth then they will simply spin in place without motion. In that case the potential energy of the falling golf ball will be converted to the rotational kinetic energy of the wheels.

Elastic Energy

(a) Mechanical Springs

There are many opportunities to employ elastic energy sources, the most common being springs and rubber bands. The properties of metal springs can be designed to match most requirements, but for MaT projects the common source of supply is the local hardware store. Springs may be purchased in a broad range of sizes. Figure 6.2 shows some examples of spring types available from manufacturers.

The first three types of springs shown in Figure 6.2 are commonly used in machinery applications. The last two indicate the types of springs that may be designed to meet special configurations and special applications. For MaT projects, the spring to be used in the project should be supplied. It is interesting to then evaluate the available energy from the spring. The Century Spring Corp. of the United States produces a range of utility springs in various sizes. A typical size is the number 302 extension spring made of hard-drawn zinc-coated steel wire with the properties shown in Table 6.3. For MaT project construction, the properties of interest are the *spring rate* and the *maximum permissible load*. Figure 6.3 illustrates the extension process and the "strain" energy stored by the process in the spring.

| Compression | Extension | Torsion | Flat clip | Wire form |

Figure 6.2 Some spring types available from Century Spring

Table 6.3 No 302 Century spring properties[6.8]					
Outside Diameter (mm)	Wire size (mm)	Working length (mm)	Initial tension (N)	Rate (N/m)	Suggetsed maximum load (N)
11.9	1.04	114.3	2.7	94.6	23

6.8 www.centuryspring.com.

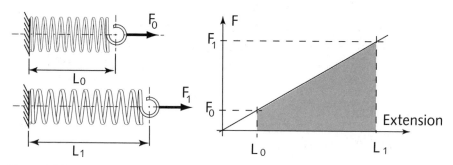

Figure 6.3 The energy stored in an extension spring. The shaded area under the extension graph is the work done on the spring during extension and this is the energy that is to be recovered if there are no internal losses in the spring

Work done on spring while expanding over its full working length is

$$W = (L_1 - L_0)(F_0 + F_1)/2$$

$$= (23/94.6 - 2.7/94.6)(2.7 + 23)/2$$

$$= 2.76 \text{ joule.}$$

If we could recover this energy in (say) one second when the spring is released, then the available power (rate of doing work) from this extension spring would be 2.76 watt. This energy must be captured by the MaT project device and transformed into some form of desired mechanical output (*e.g.*, lifting a mass to maximum height in minimum time or driving a vehicle over a track in minimum time or with maximum payload). Clearly, there will be some loss of efficiency in capturing this energy, but at least there is a basic measure of the available energy to permit planning the most desirable design of the MaT project.

There are springs available in materials other than steel, but these are not as easily acquired as the humble steel spring. Typical examples are helical springs made of plastics, or special springs of nonconducting materials. These types of springs are manufactured either for special applications to match some desired shape not easily manufactured from steel, or weight reduction or for applications where electrical insulation is imperative. Several spring manufacturers offer varieties of special springs from fibre-reinforced plastics or moulded thermoplastic materials.[6.9]

(b) Elastomers

Elastomers are a broad range of polymeric products that possess the property of elasticity (*i.e.*, the ability to regain shape after deformation). Elastomers are the basic component of most rubber products. Rubber bands used in

6.9 See, for example, Lesjöfors AB of Sweden (www.lesjoforsab.com); Advanex Inc. Japan, (www.advanex.co.jp); A useful reference for moulded springs is *Designing Plastic Parts for Assembly* (2003).

packaging and office applications are a form of elastomer. The rubber materials used as the prime movers in model vehicles (mostly airplanes) are elastomers, as are bungee cords, used in securing parcels to utility racks on motor vehicles and in bungee jumping adventures. Bungee cord is an elastomer cord covered by a fabric sheath and is available in a range of sizes and cover materials.

Polymers are chemical compounds made up of a combination of thousands of smaller molecules called monomers. Elastomers are essentially highly viscous liquids that, in their natural state, have their molecules highly disoriented and entangled. When an external force is applied to the elastomer its molecules become aligned with the direction of the force being applied. In this stretched state the elastomer's internal structure resembles the ordered crystalline structure of a rigid polymer. Once the load is removed, the elastomer's internal structure returns to its original entangled state. Elastomers have the distinctive property that they suffer virtually no change in volume as a result of deformation.

Probably the oldest and most commonly encountered elastomer product is the humble rubber band used in many packaging applications. Table 6.4 gives the range of sizes available and their designations (all dimensions are in inch units). It is a simple matter to evaluate elastic stiffness (spring rate) by direct measurement. A typical example is shown in Figure 6.4.

Figure 6.4 was obtained by direct measurement of the load-extension behaviour of a No. 63 rubber band. The elastic modulus k (the slope of the graph) was found to be 55.3 N/m and the band was found to be capable of

Table 6.4 Rubber band sizes and designations (inch units)

W	T	W	T	W	T	W	T
0.063	0.031	0.125	0.031	0.125	0.063	0.25	0.031
No.	LF	No.	LF	No.	LF	No.	LF
8	0.875	37	0.75	34	4	52	1.25
10	1.25	27	1.25	45	4.25	61	2
12	1.75	95	1.75	125	5	62	2.5
14	2	30	2	185	5.5	63	3
16	2.5	31	2.5	126	6	64	3.5
18	3	32	3	127	7	134	4
19	3.5	33	3.5	128	8	66	4.75
114	4	24	4	129	9	7005	5
115	5	—	—	—	—	—	—
117	7	—	—	—	—	—	—

W = width; T = thickness; LF = flattened length.

6.10 Source: local stationery supplier.

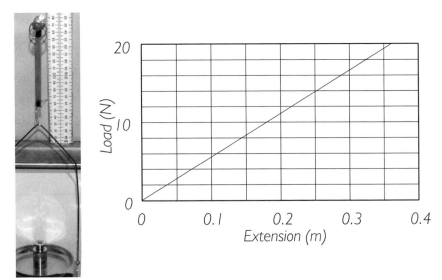

Figure 6.4 Load-extension measurement in progress. The lower paper clip was used as a pointer to measure extension. Also shown is the resulting The plot for a No. 63 Rubber band.

stretching to 0.36 m or nearly five times its unloaded length. From Table 6.4, the unloaded flattend length of the No. 63 rubber band is 3 inch (0.075 m). Therefore the energy available from the No. 63 rubber band may be estimated as

$$\text{maximum energy available} = k \times (\text{maximum extension})^2/2$$

$$= 55.3 \times (0.36 - 0.075)^2/2 = 2.25 \text{ J}.$$

With rubber bands, as with mechanical springs, the design challenge is that of being able to capture and harness this energy to do work.

(c) Compressed Air as an Elastic Energy Source

Any gas may be compressed to provide primary drive energy. Compressed air is readily available from a small portable air compressor or from a bicycle pump. To capture and harness this energy we need some means of containing the compressed gas for subsequent release to perform mechanical work. Useful containers are the 750 cc or 1 litre plastic bottles used for a variety of beverages. The energy content is a function of pressure and volume of gas contained. The energy available from a litre bottle of compressed air is of the order of 100 J, and that available from a single rubber balloon is of the order of 10 J. Moreover, for small displacements compressed air may be used as a spring with the spring constant of the order of 2 kN/m depending on the initial pressure in the air spring (refer to Appendix A1).

Electrical Energy Sources

Probably the simplest form of a portable electric energy source is available from batteries. Although photovoltaic cells and fuel cells have developed sufficiently in the last few decades to be readily available, their physical size and cost usually preclude their use in MaT projects. When batteries are specified as the prime energy source, the range of sizes and types of batteries available needs careful consideration.

Batteries are chemical energy storage systems that use electrolytic action to convert chemical energy directly into electrical energy.[6.11] The electrolyte may be any substance capable of being separated into ions with negative (*anions*) and positive (*cations*) electrical charge, respectively. When two dissimilar materials, called *electrodes,* are immersed in an electrolyte, and an electric current is permitted to pass between these electrodes, the anions will move to the electrode designated the *anode* and the cations to the electrode designated the *cathode*. The resulting chemical process is called an oxidation-reduction reaction or *redox* reaction. Figure 6.5 shows schematically the operation of an electric battery. For current to flow, the two electrodes need to be connected through a load. The flow of current follows *Ohm's Law* (after George Simon Ohm, 1789–1854), that relates current, voltage and resistance.

$$I(\text{current} \bullet \text{in} \bullet \text{amperes}) = \frac{E(\text{electric} \bullet \text{potential} \bullet \text{in volts})}{R(\text{resistance} \bullet \text{in} \bullet \text{ohms})}$$

The universally accepted symbol for current is I and the units are measured in *amperes*, named after the French physicist, Andre-Marie Ampère (1775–1836), who was a founder of the science of electrodynamics. Electric potential between the electrodes is measured in *volts*, named after Count Alessandre Giuseppe Antonio Anastasio Volta (1745–1836), who constructed an early form of the battery, known as the *voltaic pile* in 1800.

Luigi Galvani (1737–1798), a contemporary and friend of Volta, discovered the "galvanic reaction" in animal tissue when it was connected to two dissimilar metals. Galvani's first experiments used frogs and he found the legs of the frogs would twitch when touched simultaneously with two rods made of dissimilar metals. Although Galvani attributed this behaviour

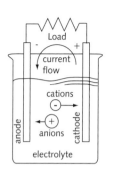

Figure 6.5 Schematic view of basic battery operation

6.11 A brief bibliography on battery development, fuel cells and photo-voltaic cells is provided in the reference list.

to some inherent property of animal tissue, Volta, an electrochemist, recognised its electromotive significance and eventually constructed the voltaic pile, the forerunner of the modern battery.

All commercial batteries work on the same principle. Two dissimilar materials (strictly speaking, they must differ in oxidation potential, commonly abbreviated as E_o value) serving as anode and cathode, are linked by a third material that serves as the electrolyte. The choice of materials for both the electrodes and electrolyte is wide, allowing for a diversity of battery technologies. The choice of electrode/electrolyte combination influences *storage density* (the amount of energy that can be stored in a given size or weight of battery) and nominal voltage output.

The two parameters that measure battery performance are electric potential (voltage) and energy content (ampere-hours). The voltage of a battery cell is determined by the materials used in it. The sum of the reduction and oxidation potentials of its electrodes is a direct measure of the unloaded voltage of the cell. For example, the discharge reaction at the positive electrode for a *lead–acid* cell is

$$PbO_2 + SO_4^{-2} + 4H^+ + 2e^- \leftrightarrows PbSO_4 + 2H_2O,$$

with a potential of 1.685 volts. The reaction at the negative electrode is

$$Pb + SO_4^{-2} \leftrightarrows PbSO_4 + 2e^-,$$

with a potential of 0.356 volts.

This gives the overall voltage of a lead–acid cell as $1.685 + 0.035 = 2.04$ volts. This value is known as the *standard electrode potential* for this type of cell. Other factors, such as the acid concentration, can also affect the voltage of a lead–acid cell. The typical open circuit voltage of commercial lead–acid cells is around 2.15 volts. Thus the voltage of any battery cell is established depending on the cell chemistry. Nickel–cadmium cells are about 1.2 volts, lead–acid cells are about 2.0 volts and lithium cells may be as high as nearly 4 volts. Cells can be connected so that their voltages accumulate. This means that lead–acid batteries with nominal voltages of 2 v, 4 v, 6 v, and so on are possible. Table 6.5 shows a brief chronology of battery development.

The two main types of batteries in use today are relatively inexpensive dry-cell nonchargable batteries, and rechargable storage batteries. Dry-cell batteries are all capable of delivering a 1.5 volt potential. When they have larger potential (9 volt, for example) they are combinations of several 1.5 volt dry-cells. Table 6. 6 lists the range of available dry-cell batteries.

Dry-cell batteries have limited shelf-life, and high temperature tends to degrade them more rapidly, although refrigeration, or in some cases freez-

Table 6.5 Brief chronology of battery development		
Date	Event	Inventor /developer
1780	Tissue contraction when touched by dissimilar metals	Luigi Galvani
1800	Voltaic pile generates useful electrical energy	Alessandro Volta
1836-1841	Copper/zinc/sulphuric acid cell – Daniell cell and various improvements	John Frederic Daniell William Robert Grove Robert Wilhelm Bunsen
1859	Lead acid storage battery	Gaston Planté
1866	Wet cell – the forerunner of the electrochemistry still employed in the cheapest form of battery in use today, the zinc–carbon dry-cell.	Georges Leclanché
1887	Patented dry cell battery	Carl Gassner

Table 6.6 (a) Cylindrical dry-cell designations and sizes[6.12]			
Type designation	Height (mm)	Diameter (mm)	Mass (gm)
AAAA	42.5	8.3	6.5
AAA	44.5	10.5	11.5
AA	50.5	14.5	23[6.13]
C	50.0	26.2	66.2
D	61.5	34.2	141.9
N	29.35	11.95	9

Table 6.6 (b) Rectangular dry-cell designations and sizes			
Type designation	Height (mm)	Width x Depth (mm)	Mass (gm)
J	48.5	33.5 x 9.2	30
9 volt	48.4	26.5 x 17.5	45.6

ing, can prolong shelf-life substantially. Storage batteries are rechargable and have a range of output characteristics. The most important of these, for manufacturers of portable electrical equipment (mobile phones and computers, for example), is *storage density*, or energy per unit mass measured in

6.12 Source : Energizer Batteries – sizes are industry standard, masses may vary between types and manufacturers.
6.13 Note the slight difference in mass from that given in Table 6.2. Published data on masses and dimensions can vary between manufacturers and direct measurement of these values is almost always a prudent approach to MaT project design.

Table 6.7 Storage battery types and storage densities[6.14]		
Cell type	Nominal voltage (volts)	Storage density (Wh/kg)
Lead–acid	2.1	30
Nickel–cadmium	1.2	40–60
Nickel–metal hydride	1.2	60–80
Circular lithium ion	3.6	90–100
Prismatic lithium ion	3.6	100–110
Polymer lithium ion	3.6	130–150

watt hours/kilogram. Table 6. 7 lists the range of storage densities available from a variety of storage batteries.

E6.5

(a) In the above description of battery power densities only the most common forms of batteries were considered. Current research into portable power technology is aimed at "environmentally friendly", super-high energy-density and nanoscale energy devices. Investigate one of these research topics and construct a table of storage densities similar to Table 6.7 for comparison.

(b) If we could harness the energy available from a storage battery and use it to launch the battery as a projectile, neglecting air resistance and other mechanical losses, estimate the maximum flight distance of the battery.

6.2 A Panoply of Dynamic MaT Projects

So far our discussion has centered on MaT projects whose specific focus of interest was the transmutation of energy. Of course, dynamic engineering devices and systems have many other, equally important, purposes. Invariably, there is a prime mover involved in the operation, and hence there is some energy conversion going on, but the main focus may be elsewhere. In preparing this section I have concentrated only on MaT projects of two specific and one more general classification (apart from the energy conversion aspects considered earlier). No doubt there are many other possibilities left unexplored here. However, the intention of this section of the book is to broaden the design and planning of dynamic MaT project opportunities available to design educators.

6.14 Crompton (2000).

Fetch and Carry MaT Projects

These are devices committed to the delivery of some payload. The payload may be already captive or it may need to be picked up from some specific location. Design challenges may be involved in the planning of the collection, capture and delivery of the payload. The selection of the specific payload may even involve some discrimination by shape, weight or colour. The choice of the project is left to the individual MaT project designer, but one typical example is offered here.

Project Across the Moat (AtM – 1980)

In this project the objective was to devise a means for delivering a payload across an approximately 1.5 metre wide, shallow reflection pool. The device was required to commence its journey on level ground and the delivery was also required to be on level ground. With this project the initial *gedanken* suggested that some form of amphibian vehicle might be used to perform the required task. Ultimately, no one could foresee the incredibly wide range of entries presented for testing, including amphibians, projectiles, mobile bridges and one outrageous and comic vehicle. The latter was a result of undue concentration on the delivery of a very large payload. Performance index for this project was designated as $W/(M \times T)$, where W was payload, T delivery time and M the mass of the vehicle excluding payload.

Figure 6.6(a) Setting up   Figure 6.6(b) Ready, set...   Figure 6.6(c) Go!

Figure 6.7 Mobile bridge   Figure 6.8 Amphibian   Figure 6.9 Projectile

The idea behind the spring-loaded vehicle, shown in Figure 6.6, was eminently simple. Given a sufficiently large spring, suitably compressed, then released, the strain energy stored in the spring would launch the long bridgelike structure across the pool and deliver a suitably large (perhaps unlimited) payload to the other side. Designers were constrained by their inability to use a self-contained launching pad for this device. They chose to embed the launching mechanism in the ground abutting the paving surrounding the pool. They also failed to consider the behaviour of a long, thin, relatively unstable vehicle under the action of a large impact load. Figures 6.6 show the sequence of events as this vehicle was launched. When the spring was released, as indicated in Figure 6.6(c), the whole vehicle simply collapsed on the spot. Almost all dynamic MaT projects have similar comic relief moments. They too add to the engaging aspects of these projects.

Figure 6.7 shows a mobile bridge with resident delivery vehicle. Figure 6.8 shows an amphibian device that needed consistent coaxing to mount the kerb surrounding the pool and then failed to navigate to the other side. Figure 6.9 shows a representative entry of the most successful types of design for this project. These vehicles launched a projectile vehicle that would deliver the payload. Most were successful in reaching the other side. However, those that did not also added to the generally appreciated comic relief of this project.

Navigationally Challenged Vehicles

Vehicles designed to navigate on some test course fall into this category. Navigational challenges may be provided by the constraints of test environments. Typical examples are:

- *DYDaF*[6.15] – Airborne vehicles in a design office with desks, doorways and light fittings;

- *PoP* – Small watercraft navigating around water drain fittings mounted at intervals in the "test track";

- *ABC* – Beverage cans navigating in a confined space of a design office.

In special examples, navigational challenges may be set by the project planners. These could form the basis of the project, where the design calls for navigating on a carefully specified track.

"How Can It Be Done?" Projects

Inherent engagement is generated in these projects from the challenge of designing a device to perform an apparently difficult task. These tasks

6.15 Refer to Chapter 7 for a description of these case examples.

include walking on water, climbing a rope to deliver an instrument, scaling a vertical wall or balancing on a wire while performing some task. Construction of bridges to scale a waterway, where the "bridge" material is confined to a small volume (a carry bag. for example), also falls into this category of project. With these types of MaT projects the guiding principles should be student safety and *doability* (*i.e.*, sufficient planning to ensure that the task is achievable).

6.3 Administration and Conduct of Dynamic MaT Projects

A major theme of this book is the underlying essence of MaT projects, their freshness and immediacy, where students and mentors struggle collectively with the design problem. Case examples are frozen in time and, at first examination, the unwary and inexperienced designer may be persuaded to accept that this neatness and linear development of ideas is a general feature of engineering design. The reality is that engineering design is an untidy, multi-stranded, problem-solving process, requiring considerable associative and lateral thinking. In the days preceding computers and calculators engineers used many graphical representations and elegant numerical short-cuts to develop their ideas. Computers tend to be seductive in their capacity to present pretty images in reporting. Calculators preclude the need for carrying out order-of-magnitude evaluations.

These issues speak to the need for using considerable caution and care in planning and administering MaT projects as an introduction to design. Perhaps a useful approach to generating a MaT project is to treat their planning stage as an exercise in design. One major departure from "real" design is that we have already identified a design "need", namely the generation of an engaging MaT project. Some desirable attributes of MaT projects are described in the following list.

1. *Primary design requirements* – Every design has some primary requirements. For dynamic MaT projects, this could be expressed as the application of some special energy source (*e.g.*, three toy balloons, or heat from a burning candle), the achievement of some specific objective (*e.g.*, jumping the longest distance), or the performance of some (perceived to be) unlikely, or difficult, task (*e.g.*, walking on water or scaling a vertical wall).

   With static MaT projects, the most frequent *primary design requirement* is that of building a structure from some unconventional engineering material (*e.g.*, newsprint or spaghetti).

2. *Subsidiary design requirements* – In "real-world" designs these requirements often remain unarticulated by the "client". Some emerge during consultations and negotiations about the proposed design solution (*i.e.*, does the proposed design meet the *primary design requirements?*). For example, in the case of a device whose primary requirement is the *removal of dirt from clothing*, a secondary (perhaps unarticulated, but assumed) requirement may be to make certain that clothing does not shrink, or drastically change colour.

With MaT projects some secondary design requirements must be articulated, whereas others will emerge during negotiations with design teams.

– *Performance index* – There is a need for easily understood performance requirements and the scales (criteria) on which the *fitness-for-purpose* of the proposed design will be measured. These requirements are usually expressed in a combination as the *performance index* (*e.g.*, payload to tare weight ratio, greatest efficiency of energy conversion or greatest speed over measured distance);

– *Environmental influences* – Testing conditions for project entries should be clearly described (e.g. indoors on carpet or outdoors on moderately rough brick paving);

– *Level of difficulty and complexity* – It has been stressed throughout the book that early thought experiments should be conducted to reveal any major inherent difficulties or complexities with the proposed project. Manufacturing requirements should be within the skill and opportunity range of all participants. The primary objective should be achievable with some relatively simple solutions. If at all possible, the "best" outcomes should contain an element of "thinking outside the square" (*e.g.*, The "string" solutions to the balsa bridge project, described in Section 5.2.1);

– *Engineering science content* – Design is the discipline where the engineering science content of engineering courses is *bedded down*, or experienced, through application. It is desirable to explore some of the engineering science content in MaT projects. Typical examples are *mechanics* (*e.g.*, rolling resistance in wheeled vehicles or conversion between potential and kinetic energies), *fluid mechanics*, (*e.g.*, air resistance or fluid behaviour in propeller- or jet-driven vehicles), *kinematics* (*e.g.*, steering mechanism and control for navigation) and *thermodynamics* (*e.g.*, heat transfer used for primary energy sources and compressed gas behaviour);

– *Safety* – An essential, though rarely articulated, requirement of MaT projects is that they should be designed to avoid hazardous situations. Complete avoidance of hazards is difficult, because minor personal injuries may be experienced even from paper cuts. Some simple rules of MaT project operation may eliminate the majority of hazards (*e.g.*, ruling out the use of high voltage or explosive devices). However, it is still possible to experience unforeseen hazards through inexperience with certain types of projects. Compressed gases and projectiles of any kind can result in hazards of this type. In the *Air Lift Engine* (*Project ALE* – 2000) MaT project the primary design requirement called for the use of compressed air in a plastic beverage bottle for raising a mass (simple energy conversion to do useful work). Some designers used electrically driven pumps to pressurise the bottles and there was some initial concern about the likely hazard of exploding plastic bottles. In the event, the actual hazard arose from the explosive behaviour of the drive systems.

Figure 6.10 shows some idea sketches of one team preparing a design for the *ALE* project. Several design teams used the compressed air to drive a piston that was used to raise a mass. In some cases, the pistons behaved as projectiles fired from a gun. Other entries used a bucket of metal parts that flew out of the bucket when the rapidly rising piston reached its end stop.

– *Presentation* – It is a sad fact of life that engineering undergraduates have been alienated from hand sketching, once the basic tool of designers. In planning MaT projects, substantial encouragement is needed to get design teams to present idea sketches. This is one specific aspect of the project where substantial mentoring may be needed.

Design restrictions and consultation – A useful rule of well-planned MaT projects is that there should be very few restrictions. They tend to seriously delimit creative opportunities. The overall spirit of the project must be teased out in carefully scheduled consultation sessions with design teams.

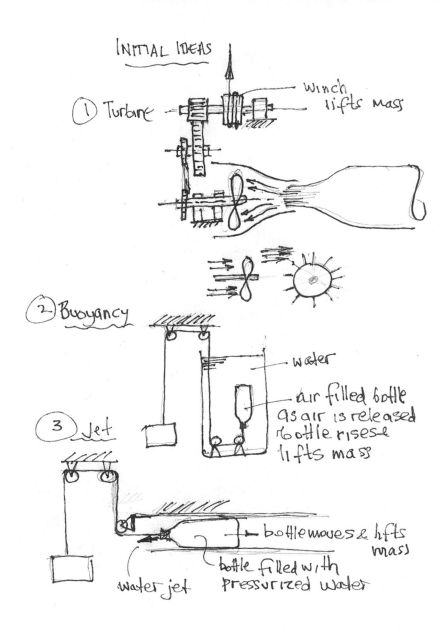

INITIAL IDEAS

① Turbine — winch lifts mass

② Buoyancy

water

air filled bottle as air is released bottle rises & lifts mass

③ Jet

bottle moves & lifts mass

bottle filled with pressurized water

water jet

Figure 6.10 Idea sketches for a lifting device using compressed air as a primary energy source

6.4 Case Studies of Dynamic MaT Projects

Chapter 7 presents detailed descriptions of some documented case examples of dynamic MaT projects, all displaying some generic similarities. In all cases, the work on the project commences with a project brief that describes the nature and intent of the project. There is always an element of competition to be considered and, if at all possible, some head-to-head competition between project submissions to make the problem solving activity personally challenging. Although this approach provides a healthy spirit of competition between student teams, it also brings with it the unhealthy fear of failure in public when the submission fails to perform. To counteract this aspect of MaT project challenges, the assessment values of the project have been carefully split between the *design report* and *prototype performance*. In addition, student teams are required:

- To prepare for interviews with the design mentoring team, and individual contributions to discussion are assessed during these interviews,

- To prepare an initial appreciation of the design problem that identifies the main elements of the design, *objectives*, *constraints* and *criteria*, and

- to present an account of the early development of design ideas, with informal sketches and descriptions in idea-logs, and a formal *design diary* of their project meetings.

In general, it is expected that students include in their reports some analysis and prediction of their expected performance in the challenge, together with a post-testing review of their predictions. These generic aspects are intended to cast the MaT project in the nature of a formal design problem and to provide experience in the formal process of designing.

The quotes at the head of this chapter are intended to point out (albeit in an abstract philosophical sense) the two key aspects of these projects. Galileo's (suspectedly *sotto voce*) statement wanted to challenge the established Christian doctrine of the time, that the Earth was the centre of the known universe. In the established, classical engineering education doctrine, educators and educational administrators find it convenient (even necessary) to bind students in a cocoon of organisation. MaT projects challenge that doctrine and place the student at the centre of importance in planning an acceptable passage through a design problem. Established, classical engineering education programmes rarely function in a Socratic dialogue format.[6.16] Yet this approach is challenging to staff (labour-intensive

---

6.16 Epstein (1970); De Simone (1968); Lotz (1995); Glickauf-Hughes and Campbell (1991); Klein (1975).

and requires considerable planning in its delivery) and students (public exposure of thinking processes, fear of failure in public). Some outcomes may be counterproductive, however, engagement with a problem and acceptance of risk in decision making are two very positive aspects of MaT project experience. Recognising personal shortcomings in knowledge and skill can, if properly managed, lead to improved learning. This book is predicated on the premise that MaT project experiences have embedded in them Socratic styles of learning.

The second quote, from the *Cornish or West Country Litany* seeks to identify the nature of these dynamic MaT projects as strange, yet familiar in the structure of formal design experience.

The case examples are arranged in a loose order of association with their focus on

(a) energy capture and conversion,

(b) ground-based vehicle design,

(c) water-based and hybrid ground/water-based transport vehicles and

(d) airborne vehicles.

In this context a "vehicle" is loosely defined as any device that is capable of carrying a payload. In addition the notion of ground, water or air as the transport medium is intended only for a defined spirit of the MaT project concerned. In almost all cases, after suitable discussion with design mentors, students were free to adopt their own interpretations of the MaT project specifications.

6.5 Experimental Evaluation of Energy Available from Various Sources

As with structural properties of materials, it is instructive to evaluate the energy available from energy sources specified in dynamic MaT projects. In what follows the two sources investigated are toy rubber balloons and mouse traps. The methods used in these experiments make use of simple equipment readily available in most households.

Energy from Toy Rubber Balloons and Mousetraps

There are two distinct forms of energy available from toy balloons. Probably the most obvious is the energy of compressed air contained in the balloon. Energy of compressed gases is partly reviewed in Appendix A1 and the ideal gas laws are used there to evaluate the energy of expansion. In the case of balloons we have the inconvenience of needing to measure the oddly shaped volume and the relatively low pressure we may be able to exhale into the balloon. The second, less obvious form of energy is that available from stretching of the ballon as an elastic. This form of energy is measured in the same way as hat available from rubber bands. Energy from mousetraps is that available from the torsion spring of the trap.

It is worthwhile to inject a note of caution (a caveat) about the results presented from direct testing (by the author). Wherever possible the tests were conducted with care and the results presented are from several tests used to verify some form of average performance. However, design teams working on MaT projects should use the results from these tests only as a guide to order-of-magnitude values. The mechanical tests reported in this book were conducted mainly as an introduction to simple experimental procedure.

EXPERIMENTAL EVALUATION OF ENERGY AVAILABLE FROM VARIOUS SOURCES	23/05/03

I. Rubber Balloon

The energy available is determined by the "enthalpy" of the gas contained in the balloon and the mass of the gas (balloon volume and pressure). Volume of the balloon is determined by using a relatively neglected theorem of Pappus that states that the volume of any rotationally symmetric solid is given by the section area rotated about the axis of symmetry multiplied by the circumference of the circle swept out by the centroid. In the case of some shapes it is significantly easier to find the section shape rotated and the centroid than the actual volume. This is certainly the case for the inflated balloon.

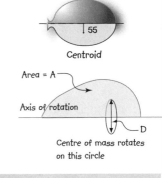

Centroid

Area = A

Axis of rotation

D

Centre of mass rotates on this circle

Locating the centroid

(a) step I – the section shape may be sketched on a piece of cardboard and cut out.

(b) step 2 – locate the centre of mass by balancing cardboard cut-out on a pin (see photo)

(c) step 3 – area A is found by weighing the cardboard cut-out on a postage balance and comparing the weight to that of a 100 x 100 mm square piece of cardboard (same caliper as the cut-out)

EXPERIMENTAL EVALUATION OF ENERGY AVAILABLE FROM VARIOUS SOURCES (continued)	23/05/03

In general this process is quite simple and quick and provides a quite accurate estimate of the volume. An alternative approach for measuring the area (in the absence of a moderately accurate postage balance) is to cut the shape from graph paper and count squares to establish the area.

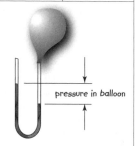

pressure in balloon

Measuring the pressure inside the balloon is also relatively easy. Using a short piece of plastic tub- Finding the pressure ing filled with water, the balloon may be inflated in the balloon and attached to one end of the tube. This simple manometer will show the height of water corresponding to the pressure in the balloon.

Using measurement method described above we have:

Volume of balloon = $\pi \times D \times A$ = 3.14 × 99 × 2.07 × 106 mm³

= 6.44 × 10⁴ mm³ = 6.4 × 10⁻³ m³

V = 0.0064 m³

Pressure in the inflated balloon was found to be 250 mm water

P = 1000 × 9.81 × 0.25 = 2453 Pa above atmospheric

The energy available from air compressed in the balloon is found by considering the expansion of the air as a constant temperature process.

P = 2453 Pa (gauge)

Atmospheric pressure p2 = 102,900 Pa

Pressure in balloon   p1 = 105,453 Pa

Inflated volume of balloon v1 = 0.0064 m³

Volume of balloon air (at atmospheric pressure)

v2 = p1 × v1/p2

>> v2 = 105,453 × 0.0064/102,900 v2 = 0.00656 m³

Hence we get work W done by the air as it expands from v1 to v2 (see figure). The work done is (approximately):

W   = (p2 + p1)(v2 − v1)/2

= [(102.9 + 105.5) × (6.56 − 6.4)]/2

= 16.7 J

W = 16.7 J

EXPERIMENTAL EVALUATION OF ENERGY AVAILABLE FROM VARIOUS SOURCES (continued)	23/05/03

If we could inflate the balloon by a further 50% (in pressure) the increase in volume of the balloon would be relatively small (recall we are using elastic energy of the balloon elastomer to contain fluid energy). Hence it is easy to show that the increase in potentially available work from such an inflated balloon would be less than 1% increase in W.

The rate at which this energy is able to do work is determined by the rate at which it can be dissipated. So, if we could deflate the balloon in (say) 1 second we could get 16.7 watts of power. The actual deflation rate (measured) is approximately 1.8 seconds. Hence the available power from a well inflated balloon is 9.3 watts.

Although this may seem a very small amount of power, even when compared to (say) an incandescent light globe, if properly captured and converted to mechanical work it can be useful. The really creative challenge is in the capture and conversion to mechanical work.

As an alternative approach to inflating the balloon we could simply use it as an elastic "spring". This alternative use of the balloon for doing work is approximately 6 J. In general this form of energy is more easily captured to do mechanical work than that available from expanding air.

2. Mousetrap

Energy available from a mouse trap spring is found by plotting the torque displacement curve for the spring and then estimating the area under the graph. In a sense this is similar to the way available energy from all elastic energy sources are estimated.

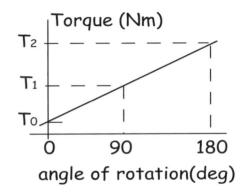

EXPERIMENTAL EVALUATION OF ENERGY AVAILABLE FROM VARIOUS SOURCES	23/05/03

T_0 (Torque needed to start spring moving)
    $= 0.147 \times 9.81 \times .045 \qquad = 0.065$ Nm
 	
T_1 (torque to rotate spring 90°)
    $= 0.353 \times 9.81 \times .045 \qquad = 0.156$ Nm
 	
T_2 Torque to rotate spring fully 180° – by extrapolation)
 	
    $= (0.156 - 0.065) + 0.156 = 0.247$ Nm
 	
>> Spring stiffness (K) $\quad = (0.156 - 0.065)/90$
$\qquad\qquad\qquad\qquad\qquad = 10^{-3}$ Nm/degree
$\qquad\qquad\qquad\qquad\qquad = (180/\pi) \times 10^{-3}$ Nm/radian
$\qquad\qquad\qquad\qquad\qquad = 0.057$ Nm/radian	

The average torque over the full 180° rotation of he spring is $$T_{average} = (K \times \pi)/2 = 0.09 \text{ Nm}$$ The energy stored in the spring when fuly rotated	W = 0.28 J
$$W_{spring} = 0.09 \times \pi = 0.28 \text{ J}$$ The trap may be sprung in a fraction of a second (estimate from some tests is about .25 second). Consequently, the maximum power available from a mousetrap is about 1 W. Not a substantial amount of energy, but may be quite useful if properly harnessed.	P = 1 W

6.6 Chapter Summary

This chapter addresses a broad range of issues that influence the design and management of dynamic MaT projects. Energy sources and their evaluation are reviewed. Some guidance is provided on the conduct and administration of dynamic MaT projects.

7
CASE EXAMPLES OF DYNAMIC MaT PROJECTS

History is the version of past events that people have decided to agree on.
Napoleon Bonaparte, *Maxims* 1804–15

History is more or less bunk. It's tradition. We don't want tradition. We want to live in the present and the only history that is worth a tinker's damn is the history we make today.
Henry Ford, 1916

When presenting case studies of projects that have documented history attached to them, one is invariably faced with the dilemma of fully describing the case or simply presenting the historical documents. Henry Fuchs on occasions described the Stanford Case Library material by noting that:

> *Documented engineering design cases are sanitised versions of reality, they always read better than they lived.*[7.1]

In what follows, I have chosen to present the documentation with some brief commentary. So that these cases should come alive for willing future participants in MaT project planning, they need to be lived through in one form or another. The mentors participating in the Melbourne projects invariably experienced a sense of *déjà vu*, not only in the organisation, but also in the reactions and performance characteristics of students.

These projects have been used at Melbourne for basic introduction to formal design procedures and teamwork. As a consequence there has always been considerable confusion at project commencement. Almost invariably, work on the project would not commence in earnest until after the first project interview. At these interviews, many teams would try to tease out the boundaries of acceptable divergence from the spirit of project specification as a means of reducing the perceived workload. This gambit was always a precursor to actually thinking about the design problem, to get the "best" performance out of the device to be designed. The next stage of idea development was (again, almost invariably) to settle on some specific approach, or idea, and to try to invent some device or process that was perceived to result in maximum performance index. This focus often involved ideas that would require fundamental research, or could be classified as elementary science fiction. Only when properly engaged with the problem did student

---

7.1 Henry Fuchs (H.O.Fuchs) was responsible for compiling the Stanford Case Library; See also Fuchs and Steidel (1973).

teams settle on "just constructing something that worked," and then go on to optimise performance. In many cases, initial problem analysis was the last recourse to be considered by student designers. Yet they were invariably informed that some early analysis can yield the best results. Wherever possible, in the case examples presented, I have indicated this "best approach" and the associated team performances.

As noted earlier, MaT projects may be categorised by the design focus intended in the project, including energy capture and delivery, as well as water-, air- or ground-based transport vehicles. Clearly, there were some inevitable overlaps between these projects. In all cases there was a need to design for energy capture and delivery, including energy conversion. For vehicles, the performance criterion was usually in the form of a measured time trial over a test track. In two cases, *Spray-can-based Towbar Pull* (*STP*) and *Mouse-trap-based Towbar Pull* (*MTP*), the direct, head-to-head performance was required. This resulted in some negative design experiences. Because vehicles in these projects were pitted against each other, in a tug-o'-war over a demarcation line, many designs concentrated on braking systems to prevent failure, rather than focusing on *"best energy conversion"* leading to success.

Invariably, our design planning team gained valuable information about setting and administering MaT projects. In this sense, the educational understanding of MaT projects became an organic, constantly changing growth experience. The following is a consolidated list of MaT projects covering a period of approximately 25 years at Melbourne.

Projects with Special Focus on Energy Capture and Conversion

- Spray can-based Towbar Pull (*STP* – 1978)
- Balloon-powered vehicle (*Huff-n-Puff* – 1982)
- One Candle Power Engine (*OCPE* – 1986/1997)
- Jumping Frog Contest (*JFC* – 1992)
- Delicate Instrument Lifter (*DIL* – 1993)
- Great Leap Forward (*GLF* – 1995)
- Air Lift Engine (*ALE* – 2000)

Projects Involving Water-based Transport

- Payloads on Pools (*PoP* – 975/1983)
- Walking on Water (*Aquaped* – 1979)
- Across the Moat (*AtM* – 1980)

- Under Water Apparatus Carrying Shell (*UWACS* – 1988)
- Mousetrap Driven Boat (MDB) 1999

Projects Using Ground Transport

- Mousetrap Driven Vehicle (*MDV* – 1985)
- Mousetrap-based Towbar Pull (*MTP* – 1989)
- A Beverage Can Roller (*ABC* – 1991)
- Golf-ball Driven Vehicle (*GDV* – 1996)
- Balloon Powered Vehicle (*BPV* – 1982)

Airborne MaT Projects

- Daring Young Designers and Flying-machines (*DYDaF* – 1973)

7.1 A Generic MaT Project Case Example

A Beverage Can Roller (ABC)

Every MaT project brief has some key generic features and these are indicated in the first fully documented case of the "A Beverage Can Roller" (*ABC*), experienced in 1991.

The Project Brief

Introduction

This project is aimed at making use of a beverage can - an essential component of modern living - in a slightly more creative way than the manufacturers intended. It will be used as a means for transporting a mechanism housed entirely inside the can. A design and prototype is required for a mechanism capable of driving an empty beverage can along a specified test track. The drive mechanism to be transported is to be housed entirely inside the can.

The Contest and Performance Index

Design teams will be supplied with a standard beverage can. Once empty, the can is to be prepared for its drive mechanism by carefully removing one end, ensuring that sufficient stiffness remains in the can end to maintain its cylindrical shape. Almost any type of drive system or energy source may be used, as long as the drive mechanism fits completely inside the empty can. For reasons of safety, explosive forms of energy supply are disallowed.

The "test track" for the beverage can roller is the floor of the design office. All cans will commence rolling from behind a designated starting line, near the northern end of the test track and continue rolling southward. The best design will be the one reaching the farthest distance (maximum roll distance *RD*) along the test track, measured in a straight line running in the North–South direction along the floor of the design office. In the event of several devices reaching similar distances (within 100 mm), a final contest will be used to establish a rank-ordered performance among the best designs. For this rank ordering the can roller reaching the best distance in 10 seconds will be deemed the winner.

If the roll distance in 10 seconds for several can rollers reaches the full length of the design office (up to the southern wall), a most unlikely condition, the roll time will be halved (to 5 seconds) and a further roll distance contest will be conducted to determine a clear winner.

Rules

- All can-driving devices must be contained entirely within the can – no protrusions or attachments outside the can will be permitted;

- The drive system may be of any form, but the use of dry-cell or storage batteries is excluded;

- The can roller must be entirely self-contained in its operation. No external attachment points are permitted for driving the can roller;

- Can rollers may be repaired after a test, but no major design alterations to the drive mechanism will be permitted;

- The finalists (if there are several similar performers) will be permitted to make significant mechanical or design alterations to their devices before the final rounds of testing, provided that all changes remain within the other rules of the contest.

Reporting

Although this is a make-and-test exercise, where performance is very important, a report of each team's design procedure is also required. The report should contain the following elements:

- *Concept development* – The way the team went about developing ideas and evaluations for the drive system, including sketches and idea-logs;

- *Downstream design issues* – Attention paid to:

  - Ease of construction and getting organised for performance testing;

  - Robustness of the design; will the device withstand the rigours of several contests?

  - Performance prediction; this will need to be reviewed after the contest to permit a pre-/post-test evaluation of the design;

– Estimate of the efficiency (distance/unit energy input) of the device;

• *Design diary* – Briefly setting out the substance of design meetings held during the project and, if possible, individual contributions by team members.

Assessment

The design report is an important part of the assessment in this project. For this reason 50% will be allocated for reporting and 50% for performance. assessed on roll distance *RD*. A factor will be allowed for constructional elegance.

Organization

Design teams will consist of two designers each and there is no restriction on choice of partnerships.

Week	Activity
1	Project brief and beverage cans available; Familiarization, partnerships advised;
2	Work on project;
3	Seminars on design strategy;
4	Testing of can roller; Post-test review, reports by 5 p.m.

Sample Entries

Initial Appreciation

Teams submitting entries to the ABC project invariably presented a formal report commencing with an *Initial Appreciation (IA)* of the design problem. The IA sets out the early evaluation of the design problem and its various subproblems as seen by the team before formal planning begins. Teams are asked to prepare, and be signed off by academic design staff, on their IA immediately after the project is handed out. Although in the final report the original IA is "tidied up", it generally provides an initial frame of reference for the design. The need to prepare an IA immediately on receiving the project, and its associated problems, provides a valuable opportunity for early guidance in planning.

The following tables set out some sample IAs presented by design teams. They are sanitised versions of the original IAs generated, but, as far as possible, illustrate the range of issues canvassed by teams in this process.

Goal: To design and construct a device or mechanism to roll a beverage can along a smooth floor for as long a distance as possible

OBJECTIVES	CRITERIA	CONSTRAINTS
High power	Torque generated (Nm)	Sufficient to rotate can i.e., overcome friction.
Travel far	Maximum distance traveled (m)	Must travel through door-ways & between desks
Travel straight	Angular deviation over 5 m (degrees)	Fit can without protrusions
Travel fast	Distance covered in 10 s (m)	
Small in size	Cross-section area (cm2), length, volume	
Lightweight	Mass (g)	< 1 kg
Reliable	Frequency of repair, life of components	Must withstand 2 tests
Inexpensive	Cost of parts ($)	< $20
Easy to manufacture	Time to produce (days)	< 30
Easy to set up	Time to set up (min)	< 10
Simple	Number of moving parts	
Robust	Drop height to fracture (m), impact strength of materials (Pa), withstand impact with wall	
Safe	Toxicity of any fumes, forces generated (N)	Nonlethal, no injury
Environmentally friendly	Toxicity of any fumes, biodegradability of materials	
Appearance (aesthetics, elegance, ingenuity)	Subjective response of 50 people (say)	

Concepts Considered

The major focus of this project was energy capture. Designers perceived (probably correctly) that specific energy[7.2] of the drive system was to be a key design issue. A clear symptom of this perception is the concentration of the idea generation phase of the design on energy sources and their various subproblems. Only few teams considered steering and navigation as a seri-

7.2 Energy per unit mass of the drive mechanism in this case.

ENERGY SOURCE	DESCRIPTION	ELIMINATION REASON
Solar power	Cover can in solar panels. Use in place of a battery	Panels expensive, circuitry too complex, panels considered protrusions
Rocket power	Use a small model rocket engine to provide momentum	Too violent to control/steer, likely to spin on spot, dangerous/expensive to test
Expanding gas propulsion	Use momentum of a gas-producing chemical reaction to turn can like a turbine	Chemicals unobtainable difficult to control reaction rate, chance of explosion, hard to control gas exit
Capacitor discharge	Use a large capacitor in place of a battery	Bending the rules, can't charge without large power source/battery, only short burst of power
Liquid motion	Mercury (heavy liquid) forced through a tube wound around can interior, provides moving "heavy side" to the can, forcing it to turn.	Heavy, toxic, could not obtain large quantities of mercury, low powered, complex equipment required, e.g., pressure vessel, low flow-rate valve
Wound rubber	Use elastic potential energy	Only 80 can rotations on rubber

HARNESSING METHOD	DESCRIPTION	ELIMINATION REASON
Clockwork	Use the gearing mechanism from a clock to control energy release from a single source	
Transmission	Use a set of gears to harness the energy from multiple sources. Allow gears to disengage when energy from each source is spent	
Sequential transmission	As above, but only allow a gear to engage when the previous gear has expended its energy	Too difficult/expensive to manufacture
Friction belts	As above but replace gears with friction belts	Gears easier to obtain, maintain

ous design issue. It may be conjectured that these were the teams that had done some field testing before presenting their "newly sanitised" idea-logs.

Morphological Analysis

This is an attribute list breakdown of the specific design project. Not all design teams prepared such a list. The ones that did so had the benefit of being able to consider a greater range of alternative designs than those teams that failed to use this approach.

ATTRIBUTE					
Energy source	Materials	Inertia source (to turn against)	Mechanism access	Construction method	Steering
Battery	Steel	Central flywheel	One end	Glue	GPS
Rubber band	Aluminium	Flywheels at both ends	Both ends	Solder	Gyroscope
Wound spring	Brass	Lead-filled rod	Surface window	Weld	Balancing
Solar	Plastic		Surface hole	Key and key-way	
Flywheel	Wood			Tie	
Clockwork	Glass			Interference fit	
Rocket	Cardboard				
Expanding gases	Foam				
Liquid motion	Ceramic				
Capacitor discharge					
Magnetic forces					

Choices (label on left side)

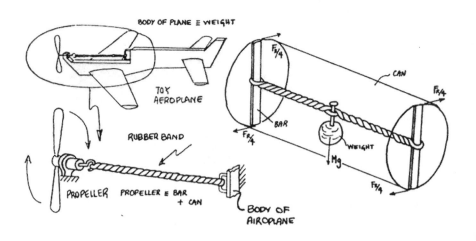

Figure 7.1 Basic ideas leading to rubber-powered drive

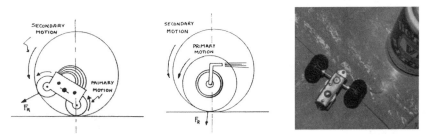

Figure 7.2(a) Clockwork   Figure 7.2(b) Jet drive   Figure 7.3 Wind-up toy drive

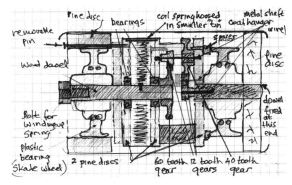

Figure 7.4 Detailed idea sketch for a spring-operated gear-drive system

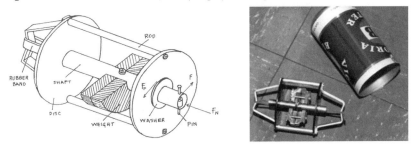

Figure 7.5 Rubber-powered inertia drive and prototype

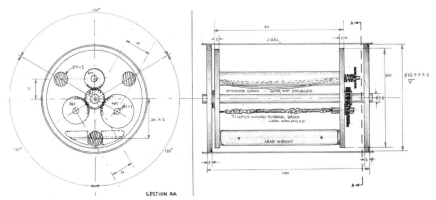

Figure 7.6 Dimensioned sketch of rubber-powered gear drive system

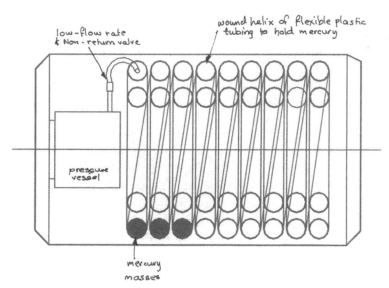

low-flow rate & Non-return valve

wound helix of flexible plastic tubing to hold mercury

pressure vessel

mercury masses

At the end, the mercury could run along a central (axial) bit of tubing where there would be a small hole (which the mercurys surface tension would be such that it wouldn't leak out, but the air could leak out ie

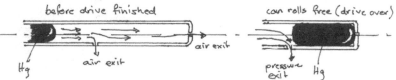

before drive finished

Hg
air exit
air exit

can rolls free (drive over)

pressure exit   Hg

Figure 7.7 Mercury drive system concept

Figure 7.8 Clockwork mechanism drive   Figure 7.9 Rubber-powered gear drive

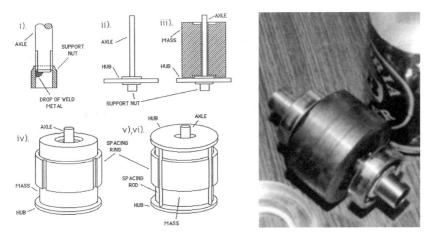

Figure 7.10 Inertia/friction drive system and prototype embodiment

Figure 7.11 The test track - view from finish line end

Figure 7.12 Rocket-powered drive   Figure 7.13 Clockwork drive

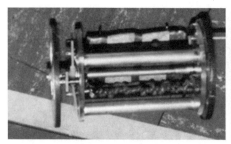

Figure 7.14 Two more rubber-powered drives, with external winders

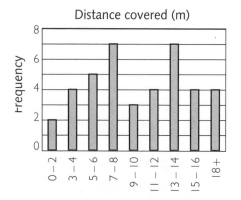
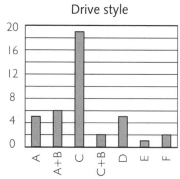

Figure7.15 Histogram of project performance   Figure 7.16 Histogram of drive style

Drive Style Legend: A = wound-up rubber band drive; B = geared drive; C = rubber band drive with counterweight; D = inertia + friction drive; E = clockwork drive (torsion spring); F = rocket motor

Concept Sketches and Photos of Entries

Figure 7.1 shows concept sketches for a rubber-powered drive presented by one team. Figures 7.2 through 7.14 show concept sketches and in some cases associated drive embodiments. Simple rubber-powered drives were the most popular. However, there were some "wayout" ideas considered by some teams, exemplified by the mercury drive idea shown in Figure 7.7. Figure 7.11 shows a view of the design office "test track" used in this project.

Performance Statistics and Review

Figure 7.15 shows the histogram of performance. Submissions ranged from simple wound rubber drives to more complex geared systems. Figure 7.16 is a histogram of the frequency and types of project submissions.

Wound-up Rubber Band Drives

This style of drive was perhaps the most easily conceptualised, at least by association with other wound rubber drives in model cars and model airplanes (refer to Figure 7.1 above). The photos of Figures 7.5, 7.6, 7.9 and 7.14 show this style of drive system. Some student teams realised early in their concept development that the wound rubber drive would need to react against something to convert the potential strain energy stored in the rubber into the kinetic energy of the rolling system. In some cases, where this issue was neglected, the can simply spun on the spot due to the low friction between the vinyl floor and the can surface. Others, who realised this feature, chose to use counterweight or geared systems to release the energy in the wound rubber more gradually. Photos in Figures 7.4, 7.9 and 7.14 show this style of drive.

Other Drive Styles

Some student teams realised that by inserting a small wound toy into the can, it could be made to roll as the toy tried to "climb up" the wall of the can. Photo of Figure 7.3 shows this style of drive submission. Clockwork spring drives, using a torsion spring, had the advantage of being capable of pre-winding and slow release of spring potential energy through the clockwork gearing. Figures 7.8 and 7.13 show examples of such an entry. Inertia/friction drive systems relied on a substantial mass wound to a high speed and the friction in the supporting bearings transmiting drive to the can. The photo of Figure 7.10 is an example of such a submission. This style of drive could certainly store substantial kinetic energy in a flywheel inertial mass. The system could be wound up by an external source such as an angle grinder or power drill. Once released the system would roll very fast indeed. In fact one such entry was the fastest of all the entries. Unfortunately the system was highly unstable, because under certain conditions the can and drive would simply stand up on one end, due to gyroscopic action, and spin like a child's toy top. Finally there were two entries that were based on rocket power, and Figure 7.12 shows a sample submission. Teams submitting such a drive were disqualified, because they clearly acted outside the project brief and, moreover, these cans did not roll but slid along the ground at high speed, mostly out of control.

Directional stability was a key factor in performance. The "test track" (the photo of Figure 7.11 shows a general view) was available for some weeks before the final contest. Once cans commenced to veer off line they were in danger of collision with desks and doorways in the room. Some student teams took considerable care in providing "friction tracks" on each end of the can, as well as balancing the load on each track, to counteract this problem. In general overall performances were best where some care and attention were paid to field-testing prior to the contest.

Lessons Learnt from the ABC Project

Teams were asked to calculate or estimate the distance their device would roll and also to estimate average velocity. Some teams attempted to carry out such analyses, but most teams simply tested their prototype and generated their estimates in this practical way. In cases where analytical estimates preceded field testing, these estimates were modified by experiences during testing. Unfortunately, most teams chose to do their tests on surfaces different from that of the test track, and eventual outcomes differed from what was expected. Navigation between desks and doorways was an unexpected difficulty encountered during final testing. Other issues involved in successful outcomes:

- Energy capture and release dominated the solution to this design problem;

- Conceptual appreciation of how stored energy may be best utilised was an important aspect of the design;

- Care and attention to detail design and early field-testing were the key elements of successful designs.

7.2 Three More Documented Case Examples

Case Example 2 – Walking on Water (Aquaped)

The Project Brief

Introduction

Figure 7.17 Common water-strider (gerris remigis)

The common water-strider (*gerris remigis*) is a small (approximately 12 mm long) insect that is capable of "walking" on, or running across the surface of ponds, by employing surface tension for support. Figure 7.17 shows a water strider in action. This project, inspired by the action of the water-strider, requires the design and construction of a device (the *aquaped*) that will allow a person (the *water-walker*) to "walk" on the surface of a body of water. The aquaped is to be propelled by the walker. No other external power supply is permitted. The action of the walker should, while providing the drive to the aquaped, mimic normal land walking gait. Each design team is required to design and construct a prototype aquaped.

The Contest and Performance Index

Two lanes (each approximately 2.5 metres wide) of the Beaurepaire swimming pool have been reserved for testing prototype aquapeds. It is intended that walkers will proceed onto the pool surface, unassisted, at the eastern end of the pool, and then proceed to walk towards the western end, where the "walk out" will also proceed unaided. The "best" design will be determined by a combination of the following factors.

- *Speed* – The walker will be timed over a measured (8 m) section of the "walking track",

- *Construction* – Simplicity of construction is an important design requirement and will be a feature used for selecting the "best" design,

- *Ingenuity and elegance* – Creative use of "aquatic technology" in achieving the desired result, coupled with elegance of design (robust mechanically sparse construction) will be the design features used for identifying the most ingenious design,

- *Cost* – This must be low. Each team must be prepared to sell their aquaped for less than A$20,

- *Durability* – The performance of the aquaped under repeated use will be considered in evaluating the "best design",

- *Confidence* – The confidence of the "walker", established by the nature of the attire worn during the test, will be taken into consideration when judging the "best design".

Rules

Other than those already stated above the following design requirements apply:

- The aquaped must not exceed 1 m in width when assembled.

- The "walker's" feet may sink into the water up to a point just below the knee. Any further sinking will disqualify the design.

Design Trophy

The two teams presenting the best and most ingenious designs will jointly become the holders of the prestigious *Creative Design Trophy* for the year.

Reporting

Reporting for this project will take the form of:

- Idea-logs, sketches, "snapshots of design team thinking" during the design process;

- A brief report (less than four A4 pages) getting it all together;

- *Post-mortem* analysis of performance following the aquaped trials, in the form of a short (5 min) seminar in which the strengths and weaknesses of each team's design should be identified, together with major design changes that might be employed in an improved design.

Organization

Each design team will consist of two designers and there is no restriction on choice of partnerships.

Assessment

Marks will be awarded as follows:

- Report and idea logs: 40%;
- Aquaped device: simplicity, effective use of materials: 20%;
- Performance: speed, confidence and ability to complete the course: 30%;
- Post mortem analysis seminar: 10%.

Sample Entries

All new, and yet untried, MaT projects have significant uncertainties associated with their eventual outcomes. Academic staff administering these projects may be able to imagine a solution, but for the participating students the outcome is an essential part of their course assessment. For this reason the "success" component of MaT projects attracts only 30% of the total marks awarded for the project, and even partial success is considered worthy of at least 15%. Of all the MaT projects offered at Melbourne, the 1979 "Aquaped" project generated the greatest concern for student performance among the academic design staff. There were 13 entries for the project, 6 of which completed the course successfully, with only 4 failing to reach the starting line of the timed test section.

The major design issues of this project were buoyancy, navigability, motive power transmission and stability. Buoyancy and motive power were the most common elements of early design focus. Most teams considered navigation and stability to be secondary considerations, to be dealt with during testing. In the event, field testing was found to be the most important element of this design. All successful teams completed significant series of field tests prior to trying their aquapeds in the pool. The following photographs and performance tables indicate the nature of entries in this project. Figure 7.18 is an idea-log page taken from the most successful student team report indicating the depth of detail with which this team investigated motive power. Figures 7.19 and 7.21 show this team's experimental approach to the project and Figure 7.20 shows a photo of one team member "walking" on the surface of the pool.

Figure 7.22 shows some detail sketches of the operation of Team 5's entry. The majority of entries used a type of hinged flap system to aid forward propulsion. Some teams tried paddles and yet others simply neglected the idea of contact friction that is an essential part of normal surface walking. Team 5 used a novel adoption of "fishtail" action used by butterfly stroke

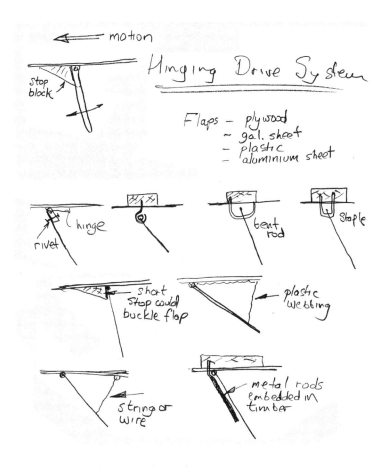

Figure 7.18 Idea sketches for aquaped propulsion system

swimmers and larger sea creatures such as whales. This system operated very well indeed and the motion of the operator's feet, when coupled with the forward motion of the device, closely mimicked natural walking. Figure 7.23 shows a member of Team 5 walking on water.

Figures 7.24 and 7.25 show photos of surfboard devices with hinged flaps in action and awaiting trial. These types of devices had the benefit of being capable of correcting navigational errors by appropriate movement of the operator's feet. Almost all floats without any forward motion assistance had serious navigational difficulties. Some examples of navigational problems are shown in Figure 7.27.

Hydrostatic stability and structural integrity were simply neglected by some teams. A pontoon float acts as a "beam" on elastic foundations, with "elasticity" provided by the buoyancy forces that support the beam. The load of the human operator acts as a point load on this beam and if the beam is structurally inadequate then it will break. Hydrostatic stability is usually

Figure 7.19 Field testing at the Art Centre

Figure 7.20 Walker in midstride

Figure 7.21 Propulsion model and one completed pontoon

encountered in early physics. The concept of metacentric height is well known to naval architects. Unfortunately those teams that neglected this concept and used buoyant but low mass floats simply fell over at the start. Some examples of these types of devices are shown in Figure 7.28. Figure 7.26 is a consolidated performance table of entries to the aquaped trials.

Design activity in this project differed markedly from that encountered in the vast majority of MaT projects. In most MaT projects field testing could be carried out in a familiar environment. With the aquaped project only few teams had access to a large water surface for field testing. The more enterprising teams made use of unusual resources such as the Melbourne Art Centre reflection pool (Team 2, Figure 7.19). Yet others made use of careful modelling. In cases where neither testing nor modelling was used the entry simply failed to perform as expected from notional, untried, ideas of stu-

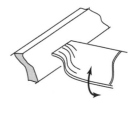

Plan view of Team 5 entry; pontoon floats and "louver" frames

Up/Down movement (steps) of walker's feet operate "louvers" for forward motion

Mylar strips on frames "fishtail" up and down providing forward motion

Figure 7.22 Operating details of the "most ingenious" entry of Team 5

Figure 7.24 Double surf-board entry

Figure 7.23 Team 5 at "take-off" and a closeup of the walking frame

Figure 7.25 Modified double surf-board entry

Team	Design Style	Time over test track (min)	Comments
1		1.21	Tubes tied together, trouble steering
2		0.18 with stocks 0.30 no stocks	Pontoons with multiple flaps - good walking action
3		0.58	Double surf-skis with masts for stability
4		0.96	Buoyant "boots" - worked well for small steps
5		0.30	Floats on rectangular frame with walking ladder flaps - good action
6		0.76	Double foam surfboards with string stabilizer - good walking action
7		0.60	Pontoons with multiple hinged flaps. Did not finish
8		Did not get to starting line	Unwieldy floating "boots"
9		3 min/3 metre - did not finish	Pontoons with extended arm paddles - poor navigation
10		Major structural failure	Tapered vee pontoons with fairing
11		Did not get to starting line	Multi-tube with flaps underneath - poor steering
12		Did not get to starting line	Inner tubes tied together with platform - completely unstable
13		Did not get to starting line	Foam flower pots combined for each float - unstable

Figure 7.26 Aquaped trials table, showing design style and performance index

Figure 7.27 Navigationally challenged entries

Figure 7.28 Entries with structural and stability problems

dents. Clearly, some teams only considered buoyancy and "hoped" that forward motion would "take care of itself." Others considered both buoyancy and propulsion but failed to consider stability and navigation. Only teams who tested their device became fully aware of the complex interactions of all four aspects of this design problem.

Lessons Learnt from the Aquaped Project

- Field testing is particularly important with MaT projects where experience with the specific design is limited or nonexistent, as was the case with the aquaped project,

- Over-concentration on the primary objectives of the project (buoyancy and forward propulsion) tends to mask some subsidiary objectives that may be critical to the success of the outcome (navigation, hydrostatic stability, structural integrity in this case),

- When setting projects that involve unusual test conditions or test tracks not easily accessible to student teams, careful instruction or mentoring is needed to alert students to the subleties of the design project and help devise ways to tackle the modelling of the problem.

Case example 3 – One Candle-power Engine (OCPE)

The Project Brief

Introduction

The British Association for the Advancement of Science (or simply BA) is launching a search for a one candle-power engine (OCPE). A prize is offered for ingenuity, efficiency and the demonstrated understanding of the scientific principles involved. The rules state that the engine may use a boiling liquid, expanding gas, bimetallic strip, thermoelectric device or anything else that will enable the heat given off by the burning wax of the candle to be turned into power.

The engine may be built with tin, brass, wood, rubber or more unusual materials, but the engine must be powerful enough to lift a small weight. If the engine is mounted on wheels, it must demonstrate an ability to travel up a gentle slope.

The candle heat source must not burn the candle at a rate greater than approximately 9 gm/second. Designers will be asked to produce evidence to establish the maximum burn rate. The main constituent of common forms of candle wax is paraffin. However, there are many different types of candles available including beeswax and specialty candles, containing colouring as well as other types of additives. Table 6.2 provides information about the heat content of paraffin, but it is expected that teams will perform their own evaluation of the heat content (specific energy) available from the candle they use in their prototype OCPE.

The Contest and Performance Index

A design and demonstration prototype of a one candle-power engine is required, eventually to be entered in the British Association OCPE contest. Performance of all engines will be determined by developed power over a one minute period. Each design team will be given three minutes to set up their OCPE for the demonstration test run. This time limit has been set by BA to ensure a level playing field for the large numbers of contestants expected to participate in this challenge.

OCPEs will be rank-ordered on performance and the one with the best demonstrated developed power will be deemed the winner on performance. If there are closely similar performance OCPEs, the best performers will be raced off against each other in a time trial to determine a clear winner on performance.

Reporting

There are two parts to the project, namely a *concept preparation* phase, and a *product demonstration* phase.

Concept Preparation

Design teams have been commissioned by a client to prepare a design concept for the *One Candle Power Engine*. The concept must pay particular attention to the following issues.

- *Simplicity* – Essential for operational and manufacturing reliability, this aspect of the design is closely associated with mechanical sparsity,

- *Ease and cost of manufacture* – This aspect of the project relates to the use of simple and well-established manufacturing methods (*e.g.*, round holes are cheaper and easier to make than square ones); The design proposed is a "one-off" device and its manufacture and assembly should not require special jigs or as yet untried construction methods,

- *Ease of setting up and maintenance (robustness)* – Only three minutes are allowed for setting up and "re-runs" may be required for the final head-to-head contest, Consequently, ease of getting organized for a demonstration run is a key feature of the design,

- *Novelty* – Special recognition will be given to devices that make use of novel forms of energy capture and transformation into useful work. This aspect of the design is still expected to attend to the previous three issues.

The *design concept* must be prepared in the form of a report with good quality sketches, indicating the overall layout dimensions of the device and its components. Clarity of presentation and reasoning to establish predicted developed power are essential. Design teams will need to convince the client that their specific chosen device is the "best" for the BA challenge.

Product Demonstration

The client expects a successful demonstration of the operation of all proposed OCPEs. Power output, elegance (related to simple application of technology, ease of setting up and robustness of operation) and ingenuity (novelty) of the design will be assessed.

Assessment

Reporting will attract 50% of the total assessment for this project. The remaining 50% will be for OCPE performance, contributions to team/client consultation sessions. A factor will be allowed for constructional elegance and ingenuity. This latter aspect of the design will be assessed by a ranked vote from all design teams, based on the photographs of the OCPEs displayed in the Design Display Cabinet.

Organization

Design teams will consist of two members each and there is no restriction on the choice of partnerships.

Sample Entries

Table 7.1 Heat content of various candles[7.3]

Candle type	Specific heat content (kJ/g)
Paraffin	16.2
Beeswax	16.8
Birthday	8.0

Typical sample entries involved sketch ideas presented as *Initial Appreciation* of the problem, where the major issues facing design teams could be resolved early through consultation with design mentors. Student teams were asked to estimate the power output and efficiency of their devices. This specific request led to considerable concern about the heat energy produced by the burning candle and any losses incurred in transforming this energy to useful work. Strange as it may seem, several student teams had difficulty in estimating the heat energy to be gained from the candle-burning process. The *burning rate* was specified, but it was recognised that there are several different types of candles that might prove to have different specific heat content. Because most readily available candles are made of paraffin wax, the energy available from them is simply a product of specific heat content of parrafin and mass rate of burning. One team estimated the heat content by direct experimentation on heating water. Their results are shown in Table 7.1.

There are several sources that list the expected value of specific energy for paraffin as 42 kJ/g.[7.3] However, this a direct calorimetric combustion value and in experiments with burning candles a severe reduction in this energy content is encountered due to losses in the heat transfer process. The values in Table 7.1 are probably more realistic values encountered when heating water directly. Of course, there are other losses to be encountered in the conversion of heat from water (say) to mechanical work. In fact, this is the specific problem with which the OCPE project teams had to grapple.

Figure 7.29 shows a collection of early concept sketches by various design teams, Figures 7.30 and 7.31 show well-developed design ideas for the OCPE. The photographs of Figures 7.32 through 7.39 show some sample prototype OCPE entries. The vacuum lift engine, shown in Figure 7.38, clearly outclassed all other contest entries, and the captive piston engine, shown in Figures 7.31 (design details) and Figure 7.39 (prototype) claimed the peer novelty and design elegance vote. Figure 7.40 shows the consolidated statistics for this project.

For design teams participating in this project, the main design issues to be considered were energy capture and transformation to mechanical work.

7.3 Source: Direct testing by one team.
7.4 Refer to published values of energy content in fuels listed in Table 6.2.

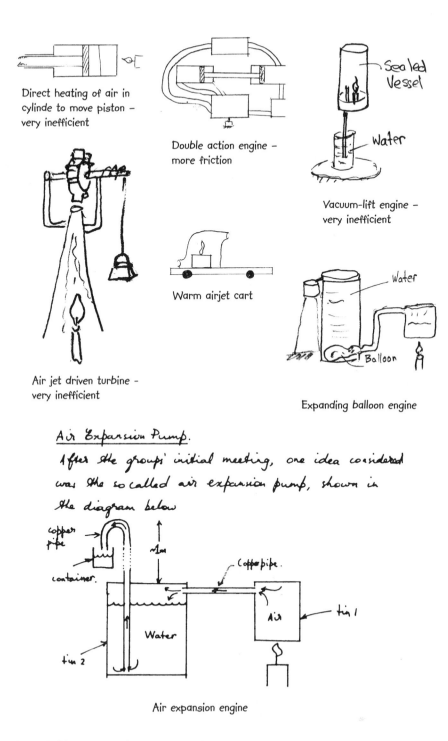

Direct heating of air in cylinde to move piston – very inefficient

Double action engine – more friction

Sealed Vessel

Water

Vacuum-lift engine – very inefficient

Warm airjet cart

Water

Balloon

Air jet driven turbine – very inefficient

Expanding balloon engine

Air Expansion Pump.

After the groups' initial meeting, one idea considered was the so called air expansion pump, shown in the diagram below

copper pipe

container.

~1m

Coffe pipe.

Air

tin 1

Water

tin 2

Air expansion engine

Figure 7.29 A collage of early concept sketches prepared by various design teams

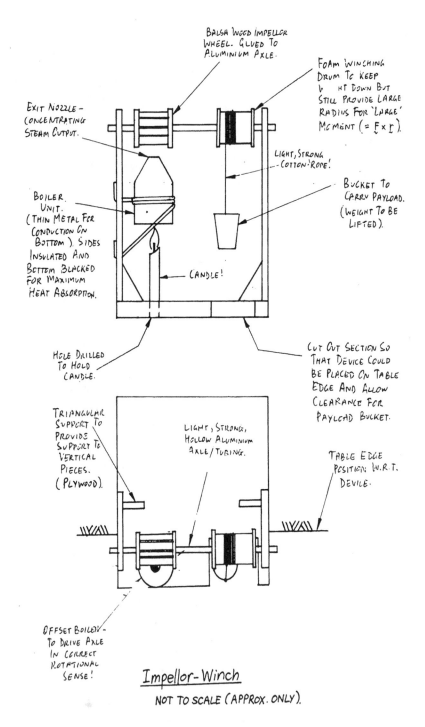

Figure 7.30 A detailed concept sketch prepared for consideration by the "client"

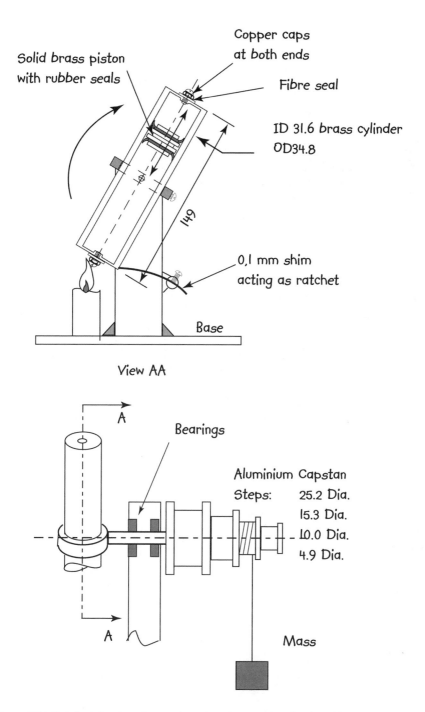

Figure 7.31 Detailed drawing of a novel captive piston engine developed by one team

Figure 7.32 Hot air and steam lift engines

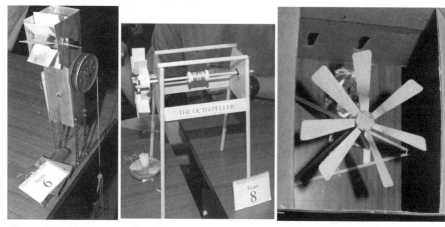

Figure 7.33 Three types of convection turbines

Figure 7.34 Three types of steam turbines

Figure 7.35 Three types of steam turbines

Figure 7.36 Bimetallic trolley

Figure 7.37 Steam engine trolley

Figure 7.38 Vacuum lift engine

Figure 7.39 Captive piston engine

Overall performances were interesting at several levels of design. Probably the simplest design idea was the hot air balloon or its variant, the vapor lift engine (usually employing toy balloons or, in some cases, condoms, when participants wished to be outrageous). One could simply boil the water or working fluid, attach a flexible diaphragm and lift a mass as the diaphragm is expanded by steam. In some cases the diaphragm was constrained to gain a more controlled larger lift, and this approach met with varying success. Teams using the hot air balloon figured that they may do better than the vapor lifters, because their approach eliminated the loss between candle heat and water. Of course, these teams did not take into consideration the relative specific heats of the two working fluids (air and water). The photos of Figure 7.32 show samples of hot air lift and vapour lift entries.

Hot air convection turbines were seen by some teams as adequate energy converters.Yet the vapour lift engines were found to be at least two orders of magnitude more efficient than these convective turbines. This type of gross error in planning speaks rather severely to these students' basic understanding of heat engine behavior. Figure 7.33 shows photos of sample convective turbine entries.

Steam turbines performed well when carefully constructed, as seen in the first two photographs of Figures 7.34. However, even the best turbine design performed at about one order of magnitude worse than the simple vapour lift engines. When poorly designed, as seen in the third photograph of Figure 7.34 (Team 39's entry), the steam turbine didn't operate at all.

Piston engines required considerable constructional skill, as well as deeper understanding of heat engine behaviour than was the case for the vapor lift devices. When carefully constructed, they fared about the same as well-designed and constructed steam turbines. When poorly designed they did not perform at all.

Figure 7.35 shows some examples of piston engine entries. None of these performed well, in spite of the considerable effort required to construct

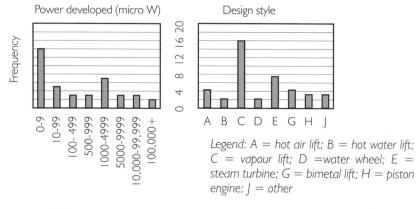

Legend: A = hot air lift; B = hot water lift; C = vapour lift; D =water wheel; E = steam turbine; G = bimetal lift; H = piston engine; J = other

Figure 7.40 OCPE statistics

them in some cases. The modified toy steam engine, shown in the central picture of Figure 7.35, performed only marginally better than the clearly self-constructed, engine of Team 17 (the third photo in Figure 7.35). Steam and water lift-driven trolleys were developed by some student teams as a means of capturing the novelty prize. Figure 7.37 shows a sample entry in this category. These types of devices almost invariably either failed to perform, or performed very poorly.

Figure 7.38 is a photo of the best performer in this MaT project, based on a vacuum lift system (refer to the sketch shown in Figure 7.29). This device performed at approximately 3.2 W, which represents a total efficiency of 8% on the calorimetric value of paraffin burning, and about 20% based on the more realistic value of paraffin wax combustion in the open air. Figure 7.39 is the photo of a novel captive piston engine (see detailed sketch in Figure 7.31) and, although low in efficiency, it was a clear winner in the novelty category. Other novel designs involved bimetallic strips and coils to raise a mass. These devices developed about 300 micro W, but they represented a class of elegant designs. Figure 7.36 shows the photo of one such device in the form of a trolley that climbed up a small ramp. Figure 7.40 sumarises the design styles adopted by OCPE design teams. Clearly, vapour lift engines were in the majority and, with a very few cases in the 0 to 9 micro W range, these vapour lifters generated performances in the 100 to 10,000 micro W range.

Another aspect of performance was the requirement to estimate power available from the burning candle as well as an estimate of overall losses in converting heat energy to mechanical work. Most teams made an attempt to estimate the heat energy available from the burning candle, some with surprising difficulty, however, very few teams attempted the estimation of heat losses. One team solved the heat equation directly using Bessel function boundary conditions for their cylindrical boiler.[7.5] Yet, in spite of their evident skills in analytical evaluation, the performance of this team's design was similar to the median performances in the vapour lift class of designs.

Lessons Learnt from the OCPE Project

- This was a particularly engaging project because it addressed some core issues in mechanical engineering, namely heat engines,

- In spite of the substantial engineering science training in the engineering course, some students exhibited remarkable weakness in understanding and application of basic thermodynamics and combustion chemistry. A positive aspect of this issue was that field testing and experimentation with the OCPE project provided invaluable guidance in clearing up some of these weaknesses.

7.5 Refer to Lienhard and Lienhard (2004).

Case Example 4 – Mechanical Frog (JFC)

The Project Brief

Introduction

In 1865 Mark Twain published a story titled "Jim Smiley and His Celebrated Jumping Frog." Sometime later the tale of this jumping frog was included in *The Adventures of Huckleberry Finn*. However, here is a brief excerpt from the earlier edition on the frog, as recounted to Twain by someone called Simon Wheeler:

> *Why, I've seen him set Dan'l Webster down here on this floor — Dan'l Webster was the name of the frog — and sing out, "Flies, Dan'l, flies!" and quicker'n you could wink, he'd spring straight up, and snake a fly off'n the counter there, and flop down on the floor again as solid as a gob of mud, and fall to scratching the side of his head with his hind foot as indifferent as if he hadn't no idea he'd been doin' any more'n any frog might do....*

> *Well, Smiley kept the beast in a little lattice box, and he used to fetch him down town sometimes and lay for a bet. One day a feller – a stranger in the camp, he was – come across him with his box, and says:*

> *"What might it be that you've got in the box?"*

> *And Smiley says, sorter indifferent like, "It might be a parrot, or it might be a canary, may be, but it ain't — it's only just a frog."*

> *And the feller took it, and looked at it careful, and turned it round this way and that, and says, "H'm — so 'tis. Well, what's he good for?"*

> *"Well," Smiley says, easy and careless, "He's good enough for one thing, I should judge – he can out–jump any frog in Calaveras county."*

Design teams have been commissioned by a client to design and construct a *mechanical frog* for an international jumping contest called the *Jim Smiley Memorial Jumping Frog Contest* (JFC).

The mechanical frog may be any device capable of executing at least one single jump. During a jump (leap, bound, hop) the frog must be airborne, with no part of it in contact with the ground. The energy needed for the jump may be derived from any source, but chemical or explosive type of fuel may not be used. The jump will commence and finish at ground level. The size of the frog is limited by a *frog hutch* (a medium size shoe-box) and the frog must be fully contained in the hutch prior to the jump contest.

The Contest

JFC contestants should prepare their frog for several test jumps in heats and finals. Each design team will be permitted up to a maximum of three test jumps. The longest jump (measured horizontally in a North–South direction along the floor of the south design office) will be deemed the team's JFC heat performance. During these heats teams may repair their frog, if it should incur some minor damage, but no substantial modification to the design of the frog is permitted. The three best performers in the JFC will be permitted to modify their design concepts as they choose and enter their original or modified design in the finals. The longest jump matching or exceeding any previous longest jumps will be declared the overall winner of the JFC. Overall performace will be determined by two performance indices, jump performance and the *frogness factor*.

- The longest jump L will be deemed the best jump performance,

- The more self-contained the mechanical frog is while airborne, the better; ingenuity and design elegance are related to the ability of the mechanical frog to mimic a real frog as closely as possible by carrying its energy source during a jump and, wherever possible, having the appearance of a real frog. This aspect of the design will be a consolidated performance index called the *frogness factor*.

The *frogness factor* will be determined by peer votes (each design team is required to submit votes) on the designs submitted. Design photos will be displayed in the Design Display Cabinet after the contests have been completed.

Rules

The following operational rules apply to this design.

- The frog must be capable of being fully contained in the hutch prior to the test jump,

- The drive system may be of any form, but chemical or explosive fuels are excluded,

- The three finalists will be permitted to make significant alterations to their devices before the final rounds of testing, provided changes remain within the other rules of the contest.

Formal consultation opportunities have been arranged between design teams and the client. As this is a "creative" design exercise, limitations on thinking outside the square should be avoided. However, it is useful design policy, as well as developing good customer relationships, to consult with the client about the design and its adherence to the spirit and rules of the JFC.

Reporting

Although this is a make-and-test exercise, where performance is very important, reporting on design procedure and planning is an equally important part of this project. Design reports should include:

- An initial appreciation of the JFC problem with idea-logs and sketches of design concepts considered, and a dimensioned sketch of the proposed "final" frog;

- An estimate of jump distance, including pre- and post-test evaluation of frog performance and efficiency (jump distance/unit energy input);
- A design diary of how the team went about planning and executing the design and, wherever possible, individual contributions by team members.

Assessment

Performance will be assessed as a combination of jump performance and peer votes attracted for frogness factor. This aspect of the project will atract 50% of the total marks available and the design report will attract the remainder of the 50% available in the project.

Organization

Design teams will consist of up to a maximum of three members and there are no restrictions on the choice of partners.

Sample Entries

The key elements of this project were the discovery of an appropriate energy source and the conversion of energy to a simple type of projectile launcher. Several design teams chose to use a catapult type of launcher where the projectile was the "frog". These types of launchers left the launch mechanism and the launch platform behind on the ground. However, in several cases these catapult frogs had the advantage of mimicking real frogs in appearance. The two other major categories of frog designs employed either

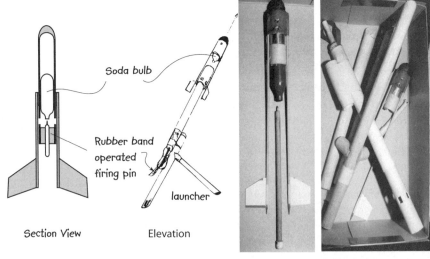

Figure 7.41 Schematic sketches for a jet-propelled glider frog submitted by Team 8

Figure 7.42 Team 8's entry in its hutch

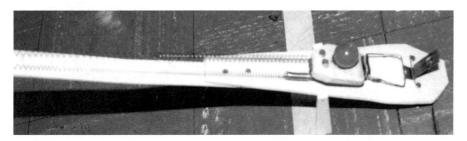

Figure 7.43 Mousetrap-launched frog

a pogo-stick approach or a compressed-air-driven jet-propulsion system. Pogo-stick designs launched a proportion of the frog that once air-borne, would carry the launch-pad section with it.

Jet-propelled frogs invariably used some form of pressurized container, with a valve, designed to discharge the pressurized fluid as a "jet" during the jump. Jet-propelled frogs clearly outclassed the majority of other designs. There were four such entries and these were treated in a separate class of designs for awarding performance assessment.

Based on jump distance the two winning design teams were Team 1, with a jump of 27.1 m and Team 8 with a jump (leap, glide) of 140 m. Team 1 used a compression spring "gun" with a steel ball projectile as their "frog". Team 8 used a jet-propelled glider as their "frog", with motive energy provided by a CO_2 soda bulb, shown in Figures 7.41 and 7.42. Team 1's entry is shown in the photo of Figure 7.46. Figures 7.44 and 7.45 show some idea sketches for frog designs, and Figure 7.43 and Figures 7.46 through 7.53 show photos of a range of entries.

In this design, *novelty* was seen by designers as a function of clever or unexpected use of materials, technology or energy sources. Constructional elegance was based on minimum use of materials and moving parts that could fail, and a robust design that performed well, and would be capable of performing well in several trials or head-to-head competitions with other frogs.

Novelty and constructional elegance (frogness-factor) were based on a subjective judgement by design teams, who were asked to vote on this by rank-ordering, in their opinion, the three most novel as well as the three most elegant designs. Photographs of all entries were displayed for this purpose and, in general, the voting seemed to be fair. Most teams tended to vote for other teams rather than themselves. This type of behaviour has been observed by Samuel and Lewis (1974), when examining project self assessment by student teams.[7.6]

Teams were asked to prepare an estimate of leap distance L for their frogs and Figure 7.54 is a sample evaluation based on ballistic projectile behav-

7.6 Samuel and Lewis (1974). See also White and Frederiksen (1998); Gladman and Lancaster (2003); Ross et al. (2003).

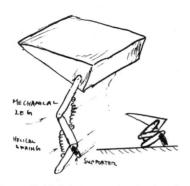

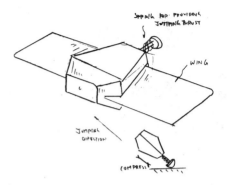

Figure 7.44 Schematic sketch plan for a frog with articulated legs

Figure 7.45 Schematic sketch plan for a "pogo-stick" frog

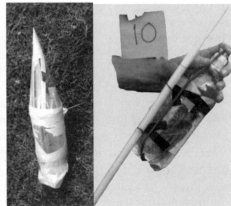

Figure 7.46 Spring "gun" launched frog of Team 1

Figure 7.47 Some more jet-propelled frogs

Figure 7.48 Catapult designs

Figure 7.49 Pogo-stick design

Figure 7.50 Articulated leg design

Figure 7.51 Pogo-stick - glider

Figure 7.52 Catapult frog

Figure 7.53 Mousetrap-launched frogs

iour. Figure 7.55 shows the consolidated performance statistics for the JFC project. Payloads were deemed to be the mass of the airborne component of the frog during a jump. This value, divided by the total mass, including the launcher, was provided by design teams as their payload-to-mass ratio. The ranges of payload-to-mass ratios ranged from 0.1 to unity. In compiling this figure the mass of the expelled fluid in jet-propelled frogs was added to the total mass.

In evaluating the frogness factor the payload–mass ratios (Rp) of frog designs were taken into acccount. However, the inclusion of Rp as a factor was an additional consideration, not initially gazetted in the project brief. Consequently frog performance was based only on the jump distance.

Lessons Learnt from the JFC Project

Although the JFC project was a relatively simple one if we consider only ballistic projectiles and catapult designs, it was surprising that only 6 out of

33 teams chose to use compressed fluid as motive power. Another surprise was that these jet-propelled frogs did not uniformly exceed the performance of elastic or spring-operated frogs. In fact Team 1 had a spring-launched design whose 27.1 m leap ($Rp = 0.13$) exceeded the majority of all but three other designs. Team 8 was identified as the overall winner on performance with a leap/flight of 140 m ($Rp = 0.35$).

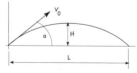

V_0 is the launch velocity; H and L are trajectory height and jump length respectively and a is the launch angle. From the initial kinetic energy of the "frog" E_0 at launch, we get (m = frog mass):

Frog launch diagram

$$E_0 = \frac{1}{2}mV_0^2 \rightarrow V_0 = \sqrt{\frac{2E_0}{m}}$$

Eo may be available from a rubber band or a helical spring. The potential energy gained at the apex of the launch trajectory is m.g.H, where g is the acceleration due to gravity. Equating the two energies (neglecting air resistance and recalling that vertical rise is due to the vertical component of the kinetic energy of the launch) we get (from Newton, neglecting air resistance):

$$H = \frac{V_0^2}{2g} \sin^2 a; \text{ and the resulting time of flight } t_F = \frac{V_0}{g} \sin a$$

The jump length L is then $\qquad L = 2t_F V_0 \cos a = \frac{V_0^2}{g} \sin(2a)$

Clearly, this is a maximum when a = 45°, which is a launch angle well known to shot-put athletes. Given that we can make the lsunch angle close to this value, then the jump length for "ballistic" frogs is $2E_0/m$.

Figure 7.54 Simple ballistic projectile physics used by some teams to optimise their design

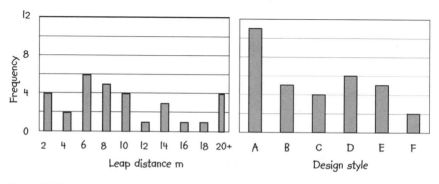

Figure 7.55 Performance statistics for the JFC project
Legend: A = rubber band catapult; B = compression spring (pogo-stick or gun); C = helical spring catapult; D = compressed fluid jet; E = mousetrap launchers; F = jet gliders

7.3 A Brief Description of Other MaT Projects

The following is a brief description of some of the remaining MaT projects from the list provided at the beginning of this chapter.

Daring Young Designers and Flying-machines (DYDaF)

This was probably the first dynamic MaT project offered at Melbourne. It arose out of one design staff member's interest in indoor gliders. The time record of airborne endurance for indoor gliders was below 1 minute. In 2000 an Australian gliding enthusiast, L.G. Surtees, set a new record of 54.8 seconds, only to be eclipsed by a pair of Japanese modellers, A. Danjo and M. Ishi, in 2003 at a new world indoor gliding record of 64 sec.

The Contest

Students were asked to build a glider, or airborne device, of mass *m,* that could remain aloft for a flying time of *t* seconds, while carrying a payload of mass *M*. There were no restrictions on materials or energy sources.

Performance Index

There was to be a maximum value of $(m \times t)/(M + m)$; novelty and constructional elegance were taken into account.

Experience

Figure 7.56 Hovercraft   Figure 7.57 Feather   Figure 7.58 Flying coolant fan

Figure 7.59 Glider 1   Figure 7.60 Glider 2   Figure 7.61 Dry ice

Most teams built winged gliders, but the more astute designers chose to build ground-effect machines (hovercrafts, see Figure 7.56) which clearly outclassed all other types of designs. There were three submissions by designers who could be described most kindly as lateral thinkers. One submission was a feather with a single lead shot attached (Figure 7.57), one was a piece of dry ice (Figure 7.61) and one, the most dangerous, a fan blade from a truck cooling pump, spun up by a power drill (Figure 7.58). None of these devices worked well, but they provided considerable amusement during testing. Figures 7.59 and 7.60 show some sample glider entries. Ultimately, ground-effect devices remained aloft, with substantial payloads, until their batteries ran out. The dry ice submission failed to remain aloft due to irregularities in the surface of the test environment. The glider submissions could not remain aloft for more than a few seconds. Their performance was seriously impeded by obstacles in the test space (desks and ceiling light fittings). Perhaps the most valuable lesson learnt from this project was the need to think through the design problem before asking inexperienced design teams to tackle it.

Payloads on Pools (PoP)

This MaT project was motivated by the desire to use an outdoor "test track" and to enforce some environmental influences on the design. The test track nominated for this project was a shallow 300 mm deep reflection pool, 55 m long and approximately 1.5 m wide on the South Lawn of the Melbourne campus. This pool was affectionately known to engineering students as "the Moat", presumably because it was an architectural feature designed to protect the Vice Chancellor's lawn.

The Contest

Design teams were supplied with a piece of pine 135 x 35 x 460 mm, for a "hull", a miniature D.C. motor (3 v/2000 rpm) and a J type[7.7] dry-cell. Designers were asked to design and build a water transport vehicle, to be raced on the Moat. The only major restriction on the design was that the final drive on these vehicles had to be transverse to the direction of motion.

Figure 7.62 A general view of "the Moat" on the South Lawn, with student teams viewing a boat race. The Moat, the brick paving alongside it and the South Lawn (to the right side of the Moat in the picture) were used in several MaT projects, where environmental influences were expected to have an impact on performance.

7.7 Refer to Table 6.6(b) for a specification of dry-cell sizes.

Figure 7.63 Dennis Miller's most ingenious paddle boat design. It went around in circles

Figure 7.64 Drive detail of the paddle boat

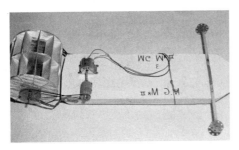

Figure 7.65 Peter McGowan's fastest boat, at an average speed of 0.5 m/s over 55 metres

Figure 7.66 Raised drains in the Moat caused some navigational hazards

Performance Index

Maximum vehicle speed over 55 metres, as well as novelty and constructional elegance were used for judging the best designs.

Experience

Two major issues influenced this design. Student designers were unused to designing waterborne vehicles, and the exclusion of propeller drives (by the restriction on output drive direction) made the design challenging. In addition, the Moat had raised drains along its centre line, to permit water circulation. These drains had a serious impact on vehicle guidance and navigation. Figures 7.63 to 7.66 show some sample entries.

The Moat featured in three more MaT projects in the years 1980 (*Across the Moat* – AtM), 1983 (slightly modified version of the *Payloads on Pools project 2* – PoP.2) and in 1999 (*Mousetrap Driven Boat* – MDB). The AtM project was designed to make use of either amphibian or airborne vehicles for carrying a payload across the Moat, and the MDB project, as the title suggests, made use of a standard mousetrap as the sole energy source in driving the boat. The only other waterborne vehicle featuring in MaT projects was the *Under Water Apparatus Carrying Shell* (UWACS – 1988), using two lanes of the Beaurepaire Swimming Pool (see the earlier *Aquaped* project) as the test track. Students were asked to design their *UWACS* devices to permit the underwater transporting of delicate instruments that may suffer damage from the water of the pool (two sample entries are shown in Figures 3.11 and 3.12).

Projects Involving Direct Head-to-head Contests (Drawbar Pull)

When a vehicle is driven by an engine through a geared transmission, the torque on the driving axle creates a force between the tyres and the road which is used to propel the vehicle. This gross force is termed the *tractive effort* and the net force; that is, the gross force minus rolling resistance is the *drawbar pull*. The relationship connecting available torque T, driving wheel radius r, transmission ratio n and rolling resistance R_r to drawbar pull D_p is

$$D_p = [(T.n/r) - R_r].$$

Two MaT projects made use of drawbar pull as a design criterion, where pairs of vehicles were matched against each other, in head-to-head combat, pulling or pushing each other over some designated line in a given time period. The process simulated a tug-o'war with wheeled vehicles. Virtually any energy source may be used for these types of projects and the two at Melbourne used Air.O.Zone spray cans in *Spray can-based Towbar Pull* (STP – 1987) and mousetraps in *Mousetrap-based Towbar Pull* (MTP – 1989).

The Contest

In the STP project design teams were supplied with a 175 g spray can of Air.O.Zone air freshener, to use as their sole energy source, in designing and constructing a vehicle, to be entered in a towbar pull contest. For the MTP project the energy source was a standard mousetrap.

Experience

Figure 7.67 A sample entry for the STP project

Figure 7.68 Two vehicles contesting the MTP project

Needless to say that after the STP project our design office reeked of various fragrances of air freshener for several weeks. Moreover, even with eco-conscious, non-CFC[7.8] based, gas carriers in the spray cans, this project erred decidedly on the environmentally unfriendly side of acceptable design projects. In essence its messiness rates alongside the "egg-drop" project with massive clean-up operations needed after the project is over. Although the MTP project is not as messy, it too has some negative elements of design project experience. As noted earlier in the introduction to this chapter, design teams tended to work more consistently on braking systems for their vehicles rather than attempting to make best use of their energy source to generate towbar pull. Figures 7.67 and 7.68 show some sample entries.

Projects Using Various Novel Forms of Energy

Designing with mousetraps has become almost a cliché in introductory design courses and even grade school and high school creative experiences. There is ample technical and practical information about the construction of devices using mousetraps.[7.9] Nevertheless, with care and perhaps a little bedevilment, mousetrap-driven devices can still provide engaging and creative challenges for undergraduate designers. In the 2004 Melbourne MaT project designers were challenged to construct a device that could start from a fixed point on level ground and travel around an object located at some variable distance and then return to as close to the starting point as possible. This project (There and Back Again – TaBA, 2004) required considerable inventive skill in using the mousetraps to both drive and steer the device around its track. Because the project intended to mimic, in some small way, a "flight around the moon", accuracy of return and travel distance were the main criteria of performance.[7.10]

In general our experience at Melbourne suggests that MaT project specifications should not restrict the source of energy to be used in the project, although some exclusions need to be applied to hazardous energy sources such as explosives. Table 7.2 lists all those MaT projects in our case list that had restricted energy sources.

Climbing and Jumping MaT Projects

There were two projects in the Melbourne list that qualify in this particular category. Both of these projects had been influenced by national challenges. In 1993 the Australian Institution of Engineers (now Engineers Australia) sponsored by Warman International (now Weir Materials Pty. Ltd.), a Sydney-based manufacturing organization, issued a national design challenge, that has since become the Annual Student Design-and-Build

---

7.8 Scientific findings identify the release of CFCs (Chloro Fluoro Carbons) into the atmosphere from refrigerants, solvents, plastic foaming and aerosol products as a major cause of ozone depletion.
7.9 See, for example, Balmer et al. (1998), Kaisinger (2002).
7.10 The TaBA project was developed and organized by John Weir at Melbourne, to whom I am indebted for permitting me to participate in judging ingenuity and constructional elegance.

Table 7.2 MaT projects with restricted energy sources

MaT project	Description	Year	Restricted energy source
OCPE	Conversion of candle heat to mechanical work	1986 1997	Burning candle .15 g/min (approximately 40 W)
Huff-n-Puff	Conversion of energy available from toy balloons to drive a vehicle	1982	Three toy balloons
STP	Conversion of energy from aerosol spray can to generate towbar pull	1987	175 g aerosol spray can
GDV	Conversion of potential energy of a golf ball to drive a vehicle	1996	A standard golf ball at 2 m above the vehicle
ALE	Conversion of energy available from a pressurised air to do mechanical work	2000	Pressurized air contained in a 1 litre plastic beverage bottle
PoP	Use a miniature D.C. motor to drive a water-borne vehicle	1975 1983	A miniature 3 v/2000 rpm motor and a J type dry cell
Metap hor	Convert the potential energy of a golf ball to operate a physics experiment	1994	A standard golf ball dropped from various heighths
MDV	Convert energy from a mousetrap to drive a vehicle	1985	One standard mousetrap
MTP	Convert energy from a mousetrap to generate towbar pull	1989	One standard mousetrap
PPL	Use the energy available from a mousetrap to launch a ping-pong ball	1990	One standard mousetrap
MDB	Convert energy from a mousetrap to drive a waterborne vehicle	1999	One standard mousetrap

Competition for the Warman Trophy.[7.11] The 1983 challenge involved the design and construction of a device that could raise a delicate instrument to a height of three metres, using a vertically suspended rope as guidance. A "delicate instrument" was simulated by a ping-pong ball in a shallow dish to be maintained at the top of the lifting device during the lift. In a slightly modified format, this project became the *Delicate Instrument Lifter - DIL.* Figure 7.69 shows idea sketches for a soda-bulb-operated DIL design. Design criteria for this project were delivery speed and successful delivery of the instrument to the top of the delivery rope. Unfortunately (for many design entries) it was the sudden stop at the end of vertical travel that tended to dislodge the "delicate instrument" from its support. In this case, controlled acceleration/deceleration coupled with maximum average speed, were the major design issues to be considered by design teams. Figures 7.71 through 7.73 show some sample DIL entries.

The *Jumping Robot* project was motivated by another national competition held in Osaka, Japan in 1994, when the Osaka Chamber of Commerce held

---

7.11 www.usq.edu.au/warman_comp/.

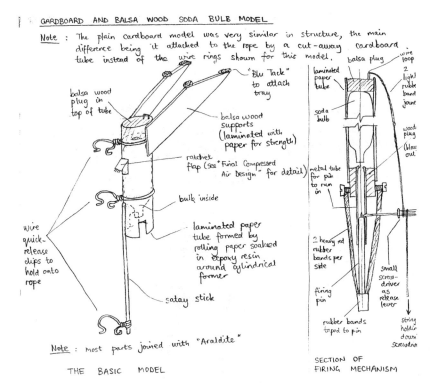

Figure 7.69 Sketch plans for a CO$_2$ soda-bulb-operated DIL

Robolympia–94, a contest of robotic construction for undergraduates in engineering. Among other things, the contest was intended to commemorate the opening of the new Kensai international airport. A Melbourne entry, a "long jump" robot, reached ninth out of 40 nations who entered the long-jump contest. The *Great Leap Forward* (GLF) MaT project in 1995 was based on the Osaka Robolympia rules for jumping robots. Figure 7.74 is a photo of the winning GLF entry. This relatively simple and inexpensive design outclassed all the *Robolympia–94* entries other than the winner with a jump of approximately 18 metres.

Experience

In the DIL project there was no restriction on the energy source to be used and most entries explored several alternative driving means for "climbing" the rope. Major types of successful entries were rope climbers typified by the photos in Figures 7.70 and 7.71, balloon lifters (photo in Figure 7.72) and extending "harmonica" systems (photo in Figure 7.73). There were several CO$_2$ soda-bulb driven entries (as typified by the idea sketch of Figure 7.69), but none were successful in completing the climb.

The GLF project had some severe restrictions on the energy source to be used in the leap. All designs had to be self-contained and could only quali-

Figure 7.70 Rope-climber designs for DIL

Figure 7.71 The winning rope climber rose at 1 m/s

Figure 7.72 A balloon design for DIL

Figure 7.73 Extending "harmonica" design for DIL

Figure 7.74 The winning design for GLF leapt 16.1 m

fy if the robot took its launching pad along for the ride during the jump. Moreover, expulsion of fluids in flight was expressly disallowed. The resulting Melbourne GLF designs were almost all of a catapult style, with the winning entry (see photo in Figure 7.74) leaping 16.1 meters, substantially exceeding the performance of the much more complex Melbourne entry for the Osaka *Robolympia–94*.

Figure 7.75 "Kangaroo" robot designed for Robolympia-94

Figure 7.76 Simpler jumping robot entry to Robolympia-94 at takeoff and in mid-flight

Figures 7.75 and 7.76 show the two proposed jumping robot entries to *Robolympia–94*. The complex, "kangaroo" style, articulated leg robot was intended to be a representative robot based on Australian native fauna. Although a nice idea in principle, the designers involved continued to pursue this design even after it was seen to be inoperable. Once released, it never left the ground due to its unstable geometry. The second, somewhat simpler design, was able to jump about 4 metres. Figure 7.76 shows this jumper in midflight.

7.4 Some Concluding Notes on Design Guidance and Engagement

In all design projects, professional or educational, the consistent need for consultation and negotiation with the "client" cannot be overstated. In MaT projects these consultations take the form of structured interviews with design teams. Not only do these interviews permit a review of progress, but they also offer an opportunity for mentoring and guidance for inexperienced designers. The following is a generic structured interview for any energy conversion focused MaT project.

Structured Interview

The following issues need to be discussed at the structured interview to be held in the second week of the project.

1. *Design concept* –
   - Identifying critical problem elements - structuring,
   - Alternative substructures considered,
   - Argument for choice of concept.

2. *Development of concept* –
   - Energy capture,
   - Losses evaluated,
   - Prediction of performance,
   - Manufacturing,
   - Reliability of operation.

3. *Current status of the design* –
   - Sketches of ideas,
   - Plans for construction.

Each team should prepare one or two A4 pages of information on their progress in the project dealing with the above issues. Performances of team members during a design interview are recorded by the "client/design mentor". An assessed value should be associated with this performance, rating up to a maximum of 10% of project assessment for substantial contributions to the progress of the design.

Guidance and Engagement

As already noted earlier, the catalogue of case examples presented in this chapter represents sanitised versions of experience. MaT projects should be designed and conducted in an encouraging and supportive environment of design guidance. ALthough they should mostly encourage the "design highs" of "aha!" experiences they should not be completely immune to the fact that design problem solving is a messy, multi-faceted and occasionally disorganized process. *Post-mortem* evaluation of the progress of design projects is invariably viewed with 20–20 hindsight. This too should be made absolutely clear to inexperienced designers. A great, though often unappreciated, benefit of a fresh new MaT project is the way established design staff and students stumble through the problem together, forming, it is hoped, a strong designer bond in the process.

Ultimately all well-designed MaT projects are learning experiences for both students and design staff who organise them. Good design planning and mentoring ensures that students engage with the problem. Carefully planned and flexible specifications ensure that the design task is seen as a challenging game against nature rather than a reactive game against the specifications (see, for example, the dry ice and feather entries in Figures 7.57 and 7.61).

7.5 Chapter Summary

This chapter presents detailed and documented case examples of dynamic MaT projects conducted at Melbourne over a period of approximately 25 years. Some care has been taken in the case study documentation to provide a uniform presentation of the various elements that may be of concern and interest to MaT project designers. The first of the cases presented is considered to be a generic example and a substantial representation of the dynamic MaT project template. Because the generation of the MaT project programme at Melbourne has always been regarded as organic and ever changing as the design team continues to encounter new experiences with almost every project, there are some slight variations from this template. In spite of these variations some care has been exercised in presenting the cases and their outcomes under a uniform set of headings including:

- The Project Brief,
- Sample Entries, and
- Lessons Learnt from the Project.

In all cases some photographs of sample entries are presented, and wherever possible a brief statistical evaluation of entries is also provided. The chapter concludes with some notes on design guidance to be used in dynamic MaT projects.

8
CONCLUDING NOTES AND SOME "CUTIEs"

> Now this not the end. It is not even the beginning of the end. But it is, perhaps, the end of the beginning.
> Winston Churchill, 1943, on the Battle of Egypt

> Le vrai voyage de la découverte consiste pas en cherchant de nouveaux paysages mais en ayant de nouveaux yeux. [The real voyage of discovery consists not in seeking new landscapes but in having new eyes.]
> Marcel Proust, À la Recherche du temps perdu [Remembrance of Things Past]

In this concluding chapter we revisit and consolidate the wide-ranging ideas loosely addressed in earlier chapters. This book has much more to do with learning to be an engineering designer than about the activity of engineering design. We have explored student motivation (engagement) and creativity as well as the broad educational framework within which these issues are addressed. What we have not yet seriously considered is the substantial human resource needed to administer and deliver the programme of learning explored in the book. There is no escaping the fact that without exception, the delivery and administration of effective engaging engineering design learning programmes is a resource intensive activity. The saving grace of participating in this activity is that although no engineer designs always, equally truly no engineer designs never.

A challenging perennial question addressed by engineering design educators is the apportionment of quality and quantity in design education. If we can think of a journey to some destination as a metaphor for the design educational experience, then quantity of education may be identified as the number of different destinations to which we can transport our students. This is an easily quantified measure and it may be considered as one of the important outcomes of the educational process. Yet as all practiced travellers know, a journey to the same destination by alternative means, or in the company of a knowledgeable guide to point out the scenic features of the journey, can deliver a vastly improved quality of the experience. The quality of the design educational experience is not easily measured in the short term. As a consequence, devoting substantial resources to delivering quality, as opposed to quantity, is a constant part of the design educator's dilemma.

An interesting teaching anecdote is about the way one teacher tried to explain the concept of infinity to a group of young primary school students by an example starting with the number 8, and dividing it by successively smaller and smaller numbers, until the result was 8/0 = infinity, represented by the symbol ∞, which looks like "eight on its side." When later the same teacher chose to test the students' understanding of this concept, he asked them to write down the value of 5/0. Many students responded by writing "five on its side." This "five-on-its side" effect is quite common in education, where the cognitive content of some concept is overshadowed by the result of an example that may be seen by students as a paradigm. Learning in engineering design can also suffer from perceived paradigms, when exposing students to design examples of our own choice. Moreover, our academic design examples are often so well defined, that they represent sanitised formulations of reality.

Engineering design is by nature, a complicated messy process. Problem statements often miss their mark of clearly identifying the design need(s) and their associated constraints and criteria. Early preparation of candidate proposals help to tease out the intricate, and often closely interlinked details of the design problem. Pahl and Beitz[8.1] refer to this aspect of design as the "design clarification" process. Educational constraints imposed on formal design experiences often tend to limit the clarification process to aspects of detail design. MaT projects offer a much more liberating experience, by allowing student designers to set their own performance limits.

It is a defining characteristic of these projects that students can't finesse answers. They have to accept responsibility for a demonstrable, tangible, and public outcome. Successful outcomes in MaT projects are invariably achieved by the combination of good synthetic thinking, sound physical understanding and careful attention to detail - all characteristics of good professional practice.[8.2]

An even more liberating learning experience (quality of learning) is available from open-ended MaT projects. The world around us has many challenging and interesting problems. We need only to identify some new and challenging problems, and to formulate the MaT project encompassing them, in a way that offers students the freedom to explore the problem and define its boundaries. These may be nonspecific problems in design that deal with basic human needs, such as food, shelter, communication and transport. Others may be simply problems that have interested us, for personal reasons. The following are some examples of problems (some have existing solutions) that have interested me as possible challenges for designers. I have coined the acronym *CUTIE – Creative Use of Technical Ideas in Engineering*, as a collective noun for these problems.

8.1 CUTIE 1: Temperature-controlled Soldering Iron

The copper tip of the soldering iron has the heat concentrated at the "pointy end". As a result, if the heating is left on permanently, as it is in some sol-

8.1 Pahl and Beitz (1996).

8.2 William Lewis founded and developed structured design programmes at the University of Melbourne between 1960 and 1998. This quote is from the foreword to this book, as well as from many personal discussions.

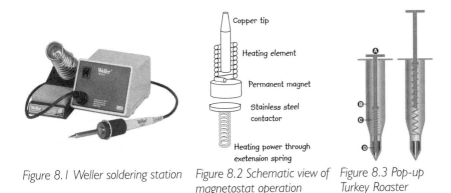

Figure 8.1 Weller soldering station   Figure 8.2 Schematic view of   Figure 8.3 Pop-up
                                        magnetostat operation   Turkey Roaster

dering irons, then the tip will burn and become pitted with copper oxide. Electronic temperature controllers may be used to counter the problem, but these controllers are expensive and still need some human intervention to set the temperature level to be maintained. A much simpler and more robust solution is provided by the arrangement shown schematically in Figure 8.2. The stainless steel contactor is magnetic at room temperatures and will maintain contact with the permanent magnet, thereby conducting power to the heating element. At around 700°C to 800°C a phase change takes place in the contactor metal and it loses its magnetic property; the contact is broken by the tension spring. By careful choice of the contactor material the temperature of the tip may be controlled by this very simple means.

P8.1 Investigate the influence of alloying additives on the magnetic properties of stainless steel. Can you think of alternative applications of the simple magnetostatic temperature controller? Could it be used at lower temperatures?

P8.2 A Pop-up Turkey Roaster (PTR) is, as the name implies, a spring-operated temperature indicator used for turkey roasting. The device is inserted into the body of the turkey and when the temperature reached inside the meat is appropriate (approximately 80°C) the PTR indicator pops up the little "Tee" signal to remove the turkey from the oven, or to turn off the heating. Investigate the way the PTR operates and suggest detailed design for such a device. Is the PTR reusable? If possible suggest alternative designs.

8.2 CUTIE 2: Galileo's Thermometer

Among other scientific experiments and discoveries, Galileo is also credited with the invention of the "thermoscope", a device for recording temperature differences (or heat), without reference to a scale of temperature. Vincenzo Viviani (1622–1703) was Galileo's assistant from 1639 to 1642. He also wrote the first officially recognised biography, *Vita di Galileo* (an histor-

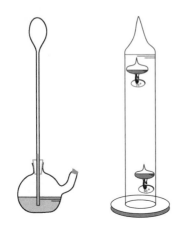

Figure 8.4 Galileo's thermoscope   Figure 8.5 Galileo's thermometer

ical account) in the form of a letter to Prince Leopoldo dei Medici, dated 29 April 1654. It was not published during the life of the author, but appeared posthumously in 1717, in the *Fasti consolari dell'Accademia Fiorentina* (Minutes of the Council of the Academy of Florence).

In Galileo's time heat and temperature were not separate concepts, and there was no consistently agreed scale on which either of these concepts could be measured. Daniel Fahrenheit (1700–1730) and Anders Celsius (1701–1744) separately invented the two currently used centigrade and Fahrenheit temperature scales respectively, in the mid eighteenth century. In *Vita di Galileo*, Viviani recounts how Galileo devised a thermoscope. Figure 8.4 shows a schematic sketch of the nineteenth century model of a thermoscope held in the *Institute and Museum of the History of Science* in Florence. This device has a sealed flask, containing some water, into which a long-necked glass retort is inverted while the bulbous end is heated. The air inside this retort expands and as it cools it draws the water from the flask into the thin tube, indicating a difference in temperature. Although Galileo did not invent a temperature scale, somehow his name became associated with thermometry. The current crop of Galileo thermometers, rightly or wrongly, continues to carry his name.

Figure 8.5 shows a currently available device called a *Galileo thermometer*. The sealed glass outer tube contains water and the small sealed bulbs contain pigmented water. These smaller bulbs are individually counterweighted to be near neutral buoyancy at various temperatures, and their tag indicates the temperature at which they are just buoyant. As the water in the outer tube heats up, buoyancy force on the smaller bulbs becomes less, and the heavier ones sink to the bottom of the thermometer. The tag on the lowest bulb remaining at the top of the jar indicates current temperature. At best, the largest forms of these colourful devices (24 inch height 12 bulbs), will indicate temperature differences of a few degrees. Nevertheless, they are attractive and colourful ornaments.

P8.3 Investigate the operation and calibration of the Galileo thermometer. If one used working fluids other than water, could the operation of such a device be improved? Explore some alternative working fluid combinations and also some alternative float materials.

P8.4 The Galileo thermometer is in a sense a "digital" device, in that it does not indicate temperature on a continuous scale. The two forces at work, in sinking and floating the indicating bubbles, are gravity and buoyancy. Investigate the possibility of using variable buoyancy forces and other types of forces (magnetic fields or mechanical springs, for example), to devise a self-contained visually appealing device to indicate temperature on a continuous scale.

8.3 CUTIE 3: Tensegrity Structures[8.3]

Richard Buckminster Fuller (1894–1983) was an American philosopher, engineer, architectural designer and a world-renowned conservationist. His interest in minimal structural design has resulted in a patented structure, called a *geodesic dome,* that is still regarded as the lightest structure capable of containing a given volume. It is also a very stable and strong structure for its size. Geodesic domes are used as communication radomes, containing and sheltering communication equipment in hostile environments, and very large storage domes for industry. Among the largest are a 165 m long x 65 m height longitudinal dome in Mexico, and a 125 m diameter hemispherical dome in Thailand.

Figure 8.6(a) shows a sketch from Fuller's geodesic dome patent, and Figure 8.6(b) is the simple underlying structure of the dome called, affectionately after Fuller, a *buckyball*. The buckyball is also the structure of an important group of new materials, named *fullerenes* also in honour of Fuller. Fullerenes are very stable forms of carbon, formed into spherelike buckyballs, when graphite is evaporated in an inert atmosphere. Usually they contain 60 or 70 carbon atoms, and a new carbon chemistry has developed around these *fullerenes*. It is possible to enclose metals and noble gases in them, to form new superconducting materials, and new organic compounds.

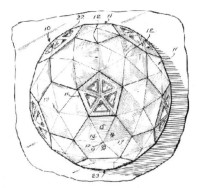

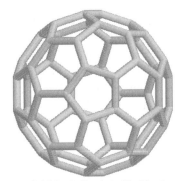

Figure 8.6(a) Sketch from Buckminster Fuller's patent of the geodesic dome

Figure 8.6(b) A "buckyball" The irregular hexagon/pentagon structure that is the basis of the geodesic dome

8.3 For further reading on Tensegrity structures see, for example, Calladine (1978); Motro (2003).

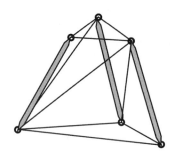

Figure 8.7 Simple T-prism tensegrity structure

Figure 8.8 Truncated dodecahedron tensegrity structure

As part of his deep interest in constructing structures with minimum materials and equipment, Fuller invented the idea of *tensegrity*, a concatenation of the terms *tension* and *integrity*. Tensegrity structures use only tension and compression members, rods and strings, in models. Figure 8.8 shows the most basic tensegrity structure, known as a *T-prism*, formed by using three rods and string tensioners to stabilise the structure. This may be constructed using timber rods and "hook eyes", tied together with string. Figure 8.9 shows a larger, more complex truncated dodecahedron tensegrity structure using 30 rods.

Tensegrity structures are used widely for constructing recreational shelters and large-scale antennae. In recent years tensegrity has received some attention as a means for constructing collapsible structures that permit deployment once released. Space antennae and large solar panel structures are two examples of such structures. These collapsible/deployable tensegrity structures are designed with spring-loaded hinge joints and, occasionally, with springs replacing some of the tension members.

P8.5 Investigate tensegrity structures and their application to collapsible/deployable antennae. Design and construct a collapsible/deployable antenna structure based on the application of several T-prism structures in series. The structure should fit into a "deploying" cylinder 200 mm diameter x 200 mm height. The best structure will be defined by the performance index $P \times H$, where P is the load carried, and H is the height at which the load is carried, measured vertically from the top of the 200 mm deploying cylinder. It is not required that any serious analysis of the structure should be performed, as that is beyond the scope of this book.[8.4]

P8.6 Investigate the construction of tensegrity polyhedra using simple materials (soda straws, rubber bands and paper clips, for example). A useful Web address for this exercise is the Web site of George Hart at www.georgehart.com.

8.4 Snelson (1973); Snelson (2002); Zeigler (1987); Zeigler (1997).

Figure 8.9 Wind generator farm

Figure 8.10 Cooling towers at a power plant

8.4 CUTIE 4: Renewable Energy Sources

Wind Farms

The Altamont Pass, one hour east of San Francisco, contains the world's largest concentration of wind turbines. The 6000 wind turbines in the Altamont pass generate 1 to 1.2 TWh (1,000,000,000 to 1,200,000,000 kWh) per year. The majority of these turbines are model 56-100 with a rotor 17.8 meters (58 feet) in diameter. Public opposition to the installation of wind farms has been raised, mainly by those concerned with their obvious visual impact on the environment. In the early days, large cooling towers used in electric power-generating plants, caused similar concerns about their visual impact. As time passes and we tend to become more attuned to changes in technology and some of these concerns abate. The photo of Figure 8.9 shows a wind farm near Dodge City, Kansas. The farm consists of 170 turbines, about 300 ft tall, with 77 ft diameter blades. They rotate at about 28 rpm and generate nearly 110 MW. This wind farm was opened in August 2001 and supplies the electrical needs of approximately 33,000 homes. Figure 8.10 shows a set of cooling towers at a conventional, fossil fuel, electric generating plant. The photos are intended to show that the visual impact of these wind generators and cooling towers are not as unappealing as the chimney stacks of older, conventional, electric power-generating plants.

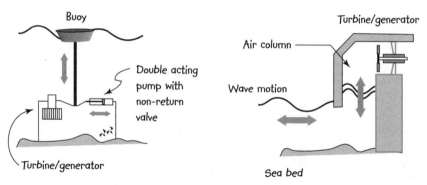

Figure 8.11 Offshore wave energy generator Figure 8.12 Shoreline wave energy generator

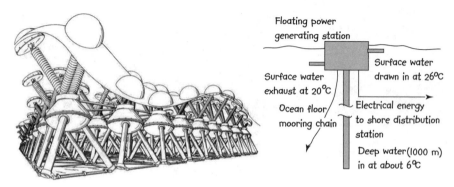

Figure 8.13 The OWECO wave motion electric power harvester

Figure 8.14 An ocean thermal energy converter

Energy from the Ocean

Figures 8.11 and 8.12 are schematic diagrams of two types of electricity generators that extract energy from wave motion. The idea of practical solutions to harvesting wave energy has been under serious investigation since the energy crisis of the early 1970s. There have been many proposals and some installations of such devices, mainly as a demonstration of their feasibility.

Figure 8.13 is a sketch of a proposal from the Ocean Wave Energy Company (OWEC; www.owec.com) of California. OWEC have been granted two patents for this device, US 4,232,230 (November 1980) and US 4,672,222 (June 1987), and a model has been tested at the company's laboratories. The patent suggests that, in one embodiment (one form) of the device, electricity may be generated directly using the up and down motion of rods inside the supporting legs of the device. Several other alternative uses of the wave motion are also identified by OWEC.

Of course there are other forms of ocean energy available as renewable energy sources. Generating technologies for deriving electrical power from the ocean include tidal power, wave power, ocean thermal energy conversion, ocean currents, ocean winds and salinity gradients. Of these, the three most well-developed technologies are tidal power, wave power and ocean thermal energy conversion. A great amount of thermal energy (heat) is stored in the world's oceans. The US Department of Energy estimates that the daily solar energy absorption of the world's oceans is equivalent to the thermal energy contained in 250 billion barrels of oil.[8.5] Figure 8.14 is a schematic sketch of an ocean thermal energy converter. The water temperature differential is used to boil salt water at very low pressures, and the "steam" is used to drive a turbogenerator.[8.6] In addition the steam loses its salinity and the desalinated water is captured from the condenser of the generating station.

P8.7 Imagine yourself as the leader of a multi-disciplinary design team at ITR (*Ideas to Reality*) Inc. A prospective client, the Renewable Energy Supply Co. (RES), has approached ITR for the design of a new type of renewable energy harvester. As a result the ITR design team has been asked to prepare candidate design proposals for feasible energy harvesters and their associated technology. From these proposals RES should be able to choose one renewable energy harvester system for further development. It is expected that ITR will support RES in building a demonstration installation of the chosen energy harvester. The ITR design team will need to prepare a status report of currently available and operating renewable energy harvesters (including wind and solar energy systems as well as ocean-based energy systems). The design team will need to rank-order the feasibility and sustainability of the three most likely candidate renewable energy harvesters. As design team leader you are asked to prepare an initial appreciation of this project with a list of tasks to be performed by the team, information resources, design objectives, constraints (including economic bounds) and performance indices.

8.5 Demonstration of Physics Principles
MaT Project Metaphor[8.7]

This project was motivated by an associated MaT project in which the author participated at Stanford in 1976. The Stanford project was itself based on some brief philosophical musings in an article written by Professor William Verplank, a respected behaviourist, who had some special concerns about language and communication in his special field of behavioural science. In his article Verplank contemplated the idea of the machine as a metaphor for living in America. In what follows some of Verplank's ideas have been translated (metamorphosed) into *physical concepts as metaphors for engineering design*.

8.5 See also www.eere.energy.gov/RE/ocean.html.
8.6 Turbo generators are close coupled turbine/electricity generator machines. See also Logan (1993).
8.7 Metaphor is generally accepted to mean an *analogy*. Its origin is from the Greek word *metafora*, meaning transport. In the broadest interpretation, a metaphor is intended to transport thought patterns between various analogies.

The Project Brief

Introduction – The Machine as a Metaphor

Language and thought have been substantially influenced by machines, and the idea of a machine, as a metaphor for everyday living, presents a strong image. Technology has had a profound influence in the development of the quality of life over the last 150 years. There have been many obvious practical influences on these developments as they progressed from the practical tinkering of some of the early settlers and the self-reliance of those on the frontier, to the expansion and exploitation of the new technology in the 19th century, spurred on by industrial mechanisation (railroads, textiles, agriculture, electricity). The rapid growth of industrial giants in the early twentieth century (automobiles, telephone, electricity generation and steel making) has led to the impersonalisation of warfare (long-range and automated weapons), the depletion of resources and the pollution of air, land and water. The quality of life, in general, has clearly improved over the last 150 years, primarily through technological progress, but with technology, we may have created as many problems as we have solved.

Machine Language, Machine Thought

There are other, less obvious, influences of technology that have to do with the way people think. The stuff of thinking is all around us; our senses bring it in. We carry with us the dominant forms, relationships and mechanisms, which we apply, amplify and modify as we face new situations, problems or people. One of the dominant realities of life for the past 150 years has been technology, and it has had a profound influence on how people think, not only about technology but about practically every aspect of their lives. The machine is one of the deepest metaphors of modern life. It is most obvious in our language.

The railroad has given us many words, including the verb *to railroad*. One might *go off the rails* or be *on the right track*, *hooked-up*, *decoupled* or *derailed*, *get up a full head of steam* or *blow off steam*. Electronics and the telephone gave us *connections*, *static* and *short-circuits*; made it possible to be *plugged-in*, *turned-on* and *in-tune*. Cybernetics has given us *input*, *output* and *feedback*. Technology has not only enriched our lives, it has enriched our language.

Underlying the assimilation of machine-language is the more profound, possibly disturbing, notion that we assimilate the characteristics of machines into our behaviour. We *process* an application, but what does it imply to *process an applicant* or to *process a patient*? *Do not fold*, *perforate* or *mutilate* applied to people suggests the metaphor of society as an information processing machine, and the citizens merely input fodder for computer processing.

Generative Metaphors

When terms from technology pass into common usage in everyday language, they become *dead* metaphors. But when they are first used, the original connection or similarity soon suggests others; the metaphor is *connected* up, *played out* and insight is gained. For example, *spaceship earth* compares the earth to a spaceship. The original connection may have been made by someone looking at the photographs of earth taken from space, seeing the comparatively small size and realising there are people on earth just as in a spaceship. But then the other connections present themselves (the metaphor generates insight): just like a spaceship, earth provides *life support*; there are problems of *energy supply* and *provisions*, the need to *recycle* with *waste-recovery* systems, which are *closed-loop*. *Spaceship earth* is a *generative metaphor*.

It is entirely possible that, when some abstract concept of engineering science or physical principle is illustrated in mechanical motion, there will be parallels or connections made between the machine-function and the situation being represented. The original connections will suggest others and insight may result. The mechanism will at least have a *start* and a *finish* and it may even *run down*. Perhaps the *Engineer's Dream Golf-ball Machine* (*EDGM*) can be a *generative metaphor*. It is equally entirely feasible that some interesting ideas might come out of attempting to illustrate abstract physical concepts in mechanical motion.

The Contest

An educational scientist client has requested some design ideas and working models for representing abstract physics concepts (the *Engineer's Dream*) in an engaging way, using some form of mechanical motion (the *Machine*). The concepts represented need not be difficult or complex, but the objective is to model them in a way that demonstrates, and, it is hoped, clarifies, the concept to young high school students.

Design teams should prepare ideas, sketches and prototype models to demonstrate the operation of their machine. Each machine will be activated by the insertion of a standard-size golf ball, following which there will be sustained motion for 30 seconds, followed by an identifiable finish. The following design criteria will be used for identifying the best *EDGM*.

- *Operating time* – It is desirable to make this as close to 30 seconds as possible. If δT is the time by which the operating time differs from 30 seconds, then δT should be minimised;

- *Self containedness* – High school students using the EDGM should be able to understand the physics principle being demonstrated without the need for explanation by any external sources. This is a subjective judgement and if *SC* is a self-containedness index on some scale then *SC* should be maximised;

- *Robustness and simplicity* – This feature of the design is a function of the number of components in the machine, its ease of operation and a clear, unambiguous representation of the physics principle it is intended to demonstrate. This feature is evaluated from subjective judgement by peer votes;

- *Degree of difficulty* – Abstract physical principles may be identified at various degrees of difficulty, depending on the number of variables involved and the nature of the relationships (equations) involved in their mathematical model. Typically, the concept of conservation of momentum is of a higher degree of difficulty than the concept of conservation of mass. This feature of the design is also based on subjective judgement by the client;

- *Elegance and ingenuity* – There is some overlap with this feature, the degree of difficulty and the design simplicity features. In general, elegance is a function of the creative way in which the concept demonstration has been developed, as well as the creative use of mechanical design in constructing the prototype model. This feature will be subjectively judged by the client in consultation with the design team.

Figure 8.15 Sound reflection in open-ended vessels

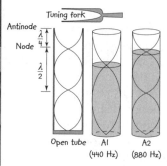

Figure 8.16 Sound vibrating in tubes closed at one end

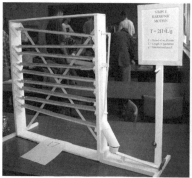

Figure 8.17 Conservation of angular momentum demonstration

This is an open-ended design problem and considerable consultation with the client will be needed before final construction of the prototype can commence.

Assessment

In this project 50% of total value will be available for the demonstration model. The remaining 50% will be awarded for the design report, detailing the planning process that led to the final choice of the physical concept to be modelled, and the design of the *EDGM* used for its representation.

Organisation

Design teams will consist of a maximum of four members, with no restriction on partnership choices.

Sample Entries

The following figures and sketches show a variety of entries submitted by design teams. Submissions ranged from the complex ideas of sympathetic resonance, diffraction of light and the conservation of angular momentum, to the relatively much simpler ideas of friction and leverage. There was no common theme that could be identified in the submissions, as was the case with those MaT projects that had a specific technical objective.

Figure 8.15 shows a device designed to demonstrate the musical scale by sequentially tapping bottles filled with varying levels of water. The bottles used in this demonstration were juice containers made of glass, 50 mm diameter and 127 mm long. The classical experiment on sound vibrating in a closed-ended tube involves a tuning fork exciting the sound waves in a vessel (refer to Figure 8.16). The vessel may be *tuned* to a specific sound *frequency*, or cycles of resonant waves per second, measured in *Hertz (*symbol *Hz)* after Herman Rudolf Hertz (1857–1894), a German physicist who was the first to produce radio waves artificially. In this demonstration the design team tuned their first bottle to 440 Hz (the sound denoted *A1* on the musical scale), followed by seven more bottles all tuned to progressively higher frequency tones in the C-major scale up to *A2* (880 Hz). As the initiating golf ball rolled down the inclined path, at regular intervals it would release spring-loaded hammers arranged to tap bottles in sequence. The basic physical principle this device was designed to demonstrate was that of the relationship between the *frequency f* and *wavelength* λ of sound, $f\lambda = c$, where c is the speed of sound in air.

The two photographs in Figure 8.17 show the front and rear views of an *EDGM* designed to demonstrate the conservation of angular momentum. The main element of this design was a wire coat-hanger *rotor*, spun up to some speed while the two masses, mounted on the rotor, were maintained in position by an elegant tension system. Once the rotor was up to speed, the masses were released and drawn in towards the centre of rotation by an elastic. As a consequence of this action, the rotation speed was seen to increase significantly. The equations of motion involved in this demonstra-

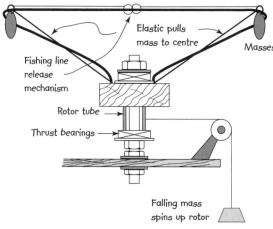

Elastic pulls
mass to centre

Masses

Fishing line
release
mechanism

Rotor tube

Thrust bearings

Falling mass
spins up rotor

Figure 8.18 Design schematic for the conservation of angular momentum EDGM

Figure 8.19 Mass release mechanism

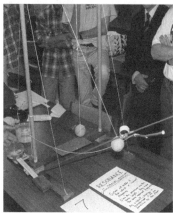

Figure 8.20 Wave diffraction

Figure 8.21 Resistance networks

Figure 8.22 Sympathetic resonance in pendula

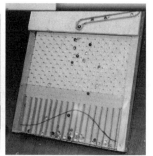

Figure 8.23 Friction

Figure 8.24 Faraday's Law demonstration

Figure 8.25 Quincunx machine of Galton

tion were shown by the display on the front of this model. The operation was initiated by a golf ball springing a mousetrap, thereby releasing a mass on a string, wound around the rotor spindle, thereby spinning up the rotor. After some carefully measured time had elapsed, a second mousetrap device released the masses held in place on the rotor. Timing was measured by a marble rolling down an inclined switchback path, seen clearly in the first photograph of Figure 8.17. Figure 8.18 shows the design schematic for this *EDGM*, and Figure 8.19 is a photograph of the mass release mechanism.

Figures 8.20 to 8.25 show photographs of some other sample entries to the Metaphor project. Figure 8.20 is a demonstration of wave diffraction by a single slit. This apparatus used a shallow water dish and a wave generator mechanism initiated by the insertion of the golf ball. Figure 8.21 shows a demonstration of parallel and series-connected resistances. The demonstration made use of coloured water flowing into different tanks via the two different resistance types. Figure 8.22 shows a demonstration of sympathetic resonance. This is an effect that is well known in music, but the effect may be also exhibited in mechanical systems, where the *oscillations* of a device (in this case a simple pendulum) can invoke *sympathetic oscillations* in another device (a second pendulum), that was formerly at rest.

Figure 8.23 shows a demonstration of friction on different types of surfaces. Although this is a relatively simple demonstration, it requires some planning to be able to explain that friction forces on a moving object depend only on the coefficient of friction and the normal force on the object being moved. Figure 8.24 is a demonstration of Faraday's Law, named after Michael Faraday (1791–1867), a pioneering experimenter in the physics of electricity and magnetism. Faraday proposed that a change in a magnetic field will induce an electric current in a conductor moving through an electromagnetic field. The conductor, made of a copper rod, was moved by the rotating wheel driven by a spring, the motion having been initiated by the insertion of the golf ball into the device. The current meter showed the change in current as the conductor moved in and out of the magnetic field. Finally, Figure 8.25 demonstrates the operation of a *quincunx*. This instrument, named after the ancient Roman coin, 5/12 (*quinque - uncia*) as well as the arrangement of five points in a square as in the figure.

$$\begin{matrix} \bullet & & \bullet \\ & \bullet & \\ \bullet & & \bullet \end{matrix}$$

The arrangement was used by Sir Francis Galton (1822–1911), an English statistician, in a device for demonstrating the statistical behaviour of engineering processes. This machine relied on an quincuncial arrangement of pins on a board (the *manufacturing process*), through which a set of marbles (the *products*) were permitted to pass without any external interference. Invariably, the pins bounced the product about, and eventually they fell through to form a normal (Gaussian) distribution, as seen in the photo.

Experience

This was an *open-ended* MaT project and it created some degree of concern among the design teams, as well as among the design mentors. A major source of concern was the need to make the demonstration mobile for a fixed time period (30 seconds). In most cases the golf ball was only used as a trigger for action and timing was done by some other means, including water pouring through a restriction and marbles rolling down inclines. In the sound generating machine the musical scale demonstration took about 11 seconds. The remaining 19 seconds were taken up by the golf ball rolling down another incline system before entering the demonstration machine. Another concern was associated with the nature of the physical concept that needed to be demonstrated. In most cases the design teams gained insight into the nature of their chosen concept through this need. In this sense the EDGM became a *generative metaphor*.

In spite of these concerns, design teams exhibited a substantial degree of enthusiasm about this MaT project. It may be conjectured that they saw it as a genuine need to be fulfilled, rather than one generated to demonstrate the design process. In most cases the resulting demonstration EDGM was not self-contained, and in many cases it required considerable interaction by the generating design team to enable the demonstration to succeed. This aspect of the project provided further insight into the nature of learning abstract physical concepts at an elementary level.

Lessons Learnt from the Metaphor Project

Open-ended MaT projects required considerably more guidance in their execution than goal-oriented MaT projects. Nevertheless, there seemed to be an unusual sense of early commitment (*engagement*) by design teams to the *Metaphor* project. For this specific reason, several open-ended MaT project examples are described earlier in this chapter (P8.1 to P8.7).

Another interesting feature of the Metaphor project was that several design teams used this opportunity to design a personally favoured laboratory experiment for demonstrating some abstract physical principle. Yet there are many such laboratory experiments used in the undergraduate engineering science streams of the course. As a consequence, there may be considerable synergy in using these engineering science experiments to focus the objectives of open-ended MaT projects.

8.6 Some Further Interesting Problems[8.8]

The urban legend about engineers is that they are rather prosaic, conservative and largely boring individuals. They are more interested in hardware than people and, in general, they prefer to contemplate the nature of engines rather than the engines of nature.

8.8 Engineering problems tend to originate from personal experiences, challenging consulting briefs and often from coffee table discussions with enthusiastic colleagues. For some of the problems in this last section I am deeply indebted to these coffee table discussions with Kenneth Brown, Kenneth Hunt, Hugh Hunt, William Lewis and John Weir.

The reality is that engineers are "doers", who are more concerned about "getting the job done" than revealing to the public at large what doing the job entails. There are few stories in which the engineer is the hero. Perhaps the best known is *The China Syndrome*[8.9], in which a lone heroic engineer is pitted against an infrastructure of management villains. The hero in that story dies and the villainous management infrastructure triumphs. Even the modest hero of the Neville Shute story *No Highway*[8.10] has difficulty in convincing others of the impending doom of fatigue failures in aircraft. It may seem that engineering stories are mostly about disasters. That perception is entirely the result of the public's general lack of awareness about good engineering that "simply works" and makes life easier for most of us. We are so accustomed to using readily available good engineering, such as dishwashers, microwave ovens, motor cars, roads, bridges, telephones and electricity, that their presence everywhere in our daily lives goes largely unnoticed.

Admittedly, in general, sex, violence and intrigue, the substance of juicy storytelling, have little in common with engineering. Yet there are many intellectually challenging problems whose elucidation and solution may reveal the real nature of what engineers love about their profession. The following is a brief list of such problems. Perhaps the list will spark off more such problems with interested readers.

The Shape of Things

As we look around us, we recognise the shape and scale of artefacts. In addition, we also recognise that the shape and scale of these artefacts are mostly governed by functional utility. Typically, the older, aerodynamically curved, shapes of refrigerators have been replaced by the more useful, greater volumetric efficiency, versions of their curved predecessors. However, there are some artefacts, large and small, whose shape, although easily recognised, does not allow such easy explanation. The following list provides some examples.

- *Cooling towers* – The shape is seen in the photo of Figure 8.10. The base is wider than the top and here is a distinct "flaring out" of the tower towards the top.

- *Fans and other fluid moving devices* – Figures 8.26 to 8.28 show three propeller blades, with distinctly different shapes, designed for three different applications. Investigate the reason for these differences in shapes.

- *Bulbous bows on ships* – Figure 8.29 shows two large ocean going vessels with bulbous bow design. Investigate the reason behind the addition of the bulbous bow to ships.

8.9 Wohl (1979). *The China Syndrome* was made into a movie by Columbia Pictures, starring Jack Lemmon, Jane Fonda and Michael Douglas.
8.10 Shute (1951). *No Highway* was made into a movie by 20th Century Fox, starring James Stewart.

Figure 8.26 Air-circulating fan blade

Figure 8.27 Propeller of an ocean-going vessel

Figure 8.28 Aircraft propeller

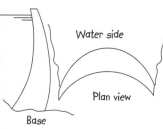

Figure 8.29 Bulbous bow on vessels in dry dock and at sea

Figure 8.30 A bell cast by Paul Revere

Water side

Plan view

Base

Figure 8.31 Eiffel tower

Figure 8.32 Schematic view of an "arch" dam

Figure 8.33 Glen Canyon dam in Arizona

- *Church bells* – Figure 8.30 shows a photo of a bell in Boston, reputedly cast by Paul Revere.[8.11] All bells have similar distinctive shapes. Investigate the reason for this specific shape;

- *The shape of some important structures* – The Eiffel Tower (Figure 8.31), an "arch" dam (Figures 8.32 and 8.33), a soccer ball (Figure

8.11 Silversmith and founder, made famous by his "ride," that marked the beginning of the American War of Independence. On the evening of April 18, 1775, Paul Revere was sent for by Dr. Joseph Warren and instructed to ride to Lexington, Massachusetts to warn Samuel Adams and John Hancock that British troops were marching to arrest them.

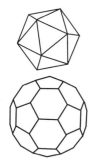

Figure 8.34 Soccer ball and icosahedron

Figure 8.35 Sydney Harbour Bridge

Figure 8.36 Flexible tape measure

8.34) and the Sydney Harbour Bridge (Figure 8.35) all have distinctive shapes. Investigate the reasons behind these shapes;

– An associated, interesting problem is that of the soccer ball. The shape is known as a truncated icosahedron, and Figure 8.34 shows the regular icosahedron, with 20 triangular faces. This figure also has 12 vertices. In the soccer ball the vertices have been truncated to 12 regular pentagons. because there are only five regular polyhedra (the Platonic polyhedra, named after the Greek philosopher and mathematician who elucidated their character), it is interesting to speculate about why it has become necessary to use a truncated regular icosahedron for the soccer ball. Moreover, it is also interesting to review why there are *only five* regular polyhedra.[8.12]

The Way Things Work

In the everyday world around us there are many simple, and sometimes surprisingly complex, devices whose workings may seem mysterious to the layperson. In many cases there are simple explanations that demistify the workings of these devices and make it clear why they do what they do, and what motivated their discovery. Some examples follow.

• *Flexible retractable tape measure* – Figure 8.36 is a photo of a typical tape measure. Note the slight inward curvature of the tape. Because these tapes may be as long as 6 or 8 metres, the torsion spring required to retract them, from the fully extended state requires careful design. The common form of design used in these tape measures is a device called a *constant force spring* motor that has almost constant torque throughout its travel. Constant force springs are a special variety of extension spring. They are tightly coiled wound bands of prehardened spring steel or stain-

8.12 See, for example: Courant and Robbins (1996); Coxeter (1973).

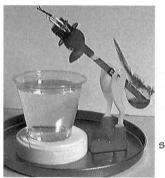

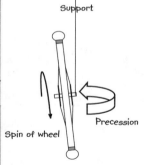

Figure 8.37 Drinking bird toy

Figure 8.38 Gyroscopic precession of a bicycl wheel

Figure 8.39 Ping-pong ball in an airjet and funnel

less steel strip, with built-in curvature, so that each turn of the strip wraps tightly on its inner neighbor. When the strip is extended (deflected), the induced stress resists the loading force in the same way as common extension springs but at a nearly constant (almost zero) rate. Investigate the operation and design of constant force extension springs.

– An associated, remarkable result is that when a flexible tape is folded over on itself the curvature of the fold is exactly the same as the inward curvature of the tape strip.[8.13]

• *Drinking bird toy* – Figure 8.37 shows a type of toy once seen in several novelty shop windows and generally known as the "drinking bird". The device is exposed to the sun and after a time it begins to rotate forward and dip its "beak" into the liquid contained in the glass. Following the dip it will continue to oscillate without dipping into the glass, with increasing amplitude of nod and eventually it will dip into the liquid again. This process continues indefinitely. Investigate the operation of this device and, with suitable, clearly stated assumptions, estimate the frequency of nodding,[8.14]

• *Gyroscopic action* – Gyroscopes and spinning tops are fascinating toys. Figure 8.38 shows schematically the gyroscopic precession action of a bicycle wheel. This is a standard laboratory experiment used to demonstrate precession. If the wheel is set to spin about its axle and then it is suspended from one side of the axle, it will not fall but it will begin to rotate (precess) about the supporting string as indicated. Investigate this action and investigate the gyroscopic behaviour of spinning tops. Identify the design variables that influence spin rate and rotation time.

8.13 See Seffen (2001); Calladine (1983).

8.14 Güémez et al. (2003).

- *Ball in a jet stream* – In old-fashioned air-rifle shooting galleries ping-pong balls were suspended in jets of water or air (Figure 8.39 shows this schematically). Once ejected by the lead pellet, they could be replaced and the ball would remain almost stably constrained in the jet. Investigate this effect. Also investigate (an associated effect) why it is difficult to dislodge a ping-pong ball from a funnel by attempting to "blow it out" (again, refer to Figure 8.39).

Limits of Performance

Performance of machinery is often of critical concern for engineers. In this context, performance is used in the generic sense of the useful output expected from artefacts or engineering systems. It is generally recognised that in manufacturing there are economies of scale. The unit cost of manufacturing a thousand artefacts is almost invariably less than the cost of manufacturing a single version. The unit cost of housing ten motor cars or pieces of farming equipment is also likely to be less than the cost of housing a single car, truck or tractor. Is it possible to extrapolate this notion of economies of scale to all engineered products or systems? How large can we build jumbo jets, or oil tankers, for example? Do these notions apply to engineering systems or organisations? For example, is the unit cost of educating an undergraduate at a university of a thousand students greater than the corresponding cost at an institute with ten times that many students? An investigation of this notion is required.

Some Not-so-clever Inventions

Being inventive is a very useful attribute of engineers. However, generating inventions and patenting them and then getting the idea to the market is a whole different ball park. There are many urban legends about the lone struggling inventor who, in the face of substantial adversity, brings some specific invention eventually to the marketplace and becomes famous and probably also very wealthy. There are some examples of such fairy-tale progress from idea to product, but in general, the time required to get an idea to market is of the order of decades (see Table 2.1). Moreover, the lone inventor rarely has the funds necessary to take the original clever idea through to a marketable product. These comments are not intended to discourage the generation of creative ideas. On the contrary, they are intended to aid and support the process of invention by advising would-be inventors to apply some discipline to the inventive process. In this section some undisciplined inventions are explored as cautionary examples.

Perpetual motion machines[8.15] – Virtually no other idea has a history as rich as perpetual motion. Probably the earliest attempts at machines working for themselves with no external energy input dates back to

---

8.15 See Ord-Hume (1977).

the application of the *Archimedes' Screw* for raising water. If the raised water could then be made to drive a water wheel, the water wheel in turn could drive the screw and so on. The idea of perpetual motion is so persuasive that, in spite of the clear objections to it from thermo-dynamic principles, inventors would try to patent such devices even as late as the early twentieth century. Eventually patent offices world wide have placed a ban on any patent that purports to be a perpetual motion machine. Instead the patent offices require a working proto-type.

Transmutation of metals[8.16] – In ancient times it was thought possible to convert base metals to precious metals by a chemical process. The device that was sought for this process was referred to as the *elixir of life* or *philosopher's stone*. Although the notion may have been a metaphor for longevity, in the first instance it was a pseudoscientific search for gold. The history of alchemy dates from the fourteenth to seventeenth centuries and has considerable implications for the way science is conducted even today.[8.17]

Antigravity devices – Levitating by magnets or by gyroscopic means has fascinated inventors for some time. This fascination may have been generated by a less than full understanding of how these two effects interact with the Earth's gravity field. Unfortunately this example of pseudoscience has influenced otherwise respectable scientific minds. The result is that the community at large has some conceptual diffi-culty in evaluating the veracity of such effects.

"What goes up must come down. Well, maybe not." *According to Popular Mechanics 3 January 1998,*

A British scientist said yesterday he is on the threshold of inventing an antigravity motor that could fly a manned spaceship to the stars using nuclear fuel the size of a pea. E. L. [name removed by author], professor… at London's Imperial College of Science and Technology, said the motor is based on the gyroscope, a rapidly spinning top that defies gravity. Gyroscopes already are used to guide spaceships. "The motor is not easy to explain. If it was, others would have tried to produce one by now," said L., who described himself as an astro engineer.

Ball operated clocks – Clocks with ball operated escapement mecha-nisms grew out of the idea of perpetual motion. The notion of a clock that would never need winding was an attractive motive behind this fallacious idea. Probably the best known example is that of Sir William Congreve (1772–1828), an inventive military engineer. He patented his rolling ball clock in 1808 (refer to Figure 8.40). Although

8.16 A useful reference on alchemy is Newman (2002).
8.17 Park, R. (2000); See also Weber (1973).

Figure 8.40 Congreve Clock

the Congreve clock is an attractive eye catcher, it is very inaccurate, mainly due to the friction between the rolling ball and its track. Congreve also tried (unsuccessfully) to take out a patent for a perpetual motion machine.[8.18]

Finally it is worth noting that some very odd problems seem to keep exercising the minds of inventive people and that is probably as it should be. An interesting compendium of odd, almost absurd patented devices is available at *http://www.totallyabsurd.com*.

Creative Ideas and Patents

Patents and creative ideas can become soft targets for scorn and occasional disrespect. A useful notion for engineering students faced with the prospect of creative behaviour is that no question or idea is totally ridiculous. Every invention, however silly it may seem, is the result of someone's perceived or real need and its associated thought process. In fact, it is a useful exercise in creative thinking to explore the world of patents and patenting, even if at a relatively elementary level. Several patent offices provide searchable data bases and the United States Patent and Trademark Office (USPTO) has its Web address at *http://www.uspto.gov/patft/index.html*. This site features a search by patent number and the following patents were found there.

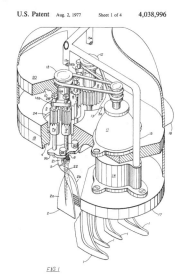

Figure 8.41 Automatic hair braider (US 4,038,996)

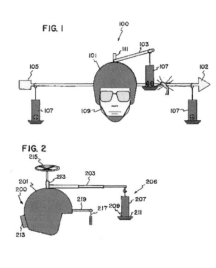

Figure 8.42 Bird feeding helmet (US 5,996,127)

8.18 A. Hart-Davies (1999).

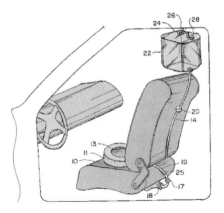

Figure 8.43 Greenhouse helmet (US
4,605,000)

Figure 8.44 Automobile urinal (US
4,344,424)

Figure 8.41 is of an invention intended to permit mechanical braiding of
hair. Figure 8.42 is an invention for observing birds in general while they are
feeding. However, the specific aim of the bird feeding helmet is to observe
humming birds. Figure 8.43 is of an invention intended to help athletes
breathe more oxygen while in training. The plants woud breathe in CO_2
exhaled by the athlete and breathe out oxygen in the close proximity of the
wearer's head. Figure 8.44 is of an automobile toilet with flushing capability.

Patents describe the invention and make claims about the invention's purposes. The intention of the description of the patent is to describe the
invention in such embracing terms that it should make it difficult for anyone to infringe the patent. For exampe, the descriptive statement for the
bird feeding helmet is:

*A device for feeding and observing flying animals comprising a hat, a support
mounted on the hat and extending outward from the hat, and a feeder mounted on
the support. When flying animals feed from the feeders, a person wearing the hat
may observe them from a short distance. The device may comprise a helmet with
three poles mounted on it and extending outward from the helmet, and a feeder
hanging from each of the poles. A variety of flying animals, including butterflies,
hummingbirds, and other small birds, may be observed with the device.*

The main body of a patent is devoted to its claims and diagrams. In general, Claim 1 in the patent is the most important and it is usually this specific claim that the inventor may need to defend in a court of law against any
infringement. For example, United States Patent 4,344,424 granted in
August 1982 for an *Anti-eating face mask* is described as follows:

*An anti-eating face mask which includes a cup-shaped member conforming to the
shape of the mouth and chin area of the user, together with a hoop member and
straps detachably engageable with a user's head for mounting the cup-shaped mem-*

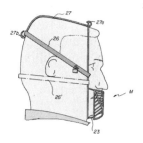

Figure 8.45 Anti-eating
face mask invention (US
4,344,424)

ber in overlying relationship with the user's mouth and chin area under the nose thereby preventing the ingestion of food by the user.

The claims for this invention are:

1. A face mask for preventing the introduction of substances into the mouth of the wearer comprising in combination;

a cup shaped member conforming generally to the shape of the mouth and chin area of the wearer's face below the nose, said cup shaped member formed of rigid material with openings to allow breathing therethrough,

means for mounting said cup shaped member over the mouth including plural straps extending from said cup over the head of the wearer,

one of said straps provided with a means for separation to allow for placement and removal of said face mask, and a lock at said separation means to prevent removal of said face mask, said mounting means further including a hoop member of rigid material adapted to extend over the user's head and chin and a flexible strap having end portions connected to said hoop member for extending around the back of the user's head and a strap member adapted to extend over the top of the user's head, said strap member being connected at opposite ends to said hoop member and to said strap intermediate the ends of said strap, said separation means including a staple adjacent one end portion of said strap, and an aperture in said strap one end portion in longitudinally spaced relationship with said staple, said strap one end portion arranged to be looped around said hoop member for insertion of said staple in said aperture and said lock engageable with said staple extending through said aperture for locking said strap one end portion to said hoop member, wherein said hoop member is attached to said cup shaped member by plural rods extending therebetween, two of said rods attached to opposed sides of said cup shaped member, a third said rod attached to a lowermost portion of said hoop member.

2. A face mask in accordance with claim 1 wherein said cup-shaped member comprises a screen formed of a plurality of wires, said wires in said screen defining said openings therebetween.

The anti-eating face mask invention is shown in Figure 8.45. For legal reasons the patent text is written in *legalese*, a language claiming to be precise, brooking no ambiguity, but unfortunately this language is also quite reader-unfriendly, as are many other legal documents. It is an interesting and challenging intellectual exercise to find ways of getting around the claims of the invention without legally infringing the original patent. Some manufacturers use this approach to "reverse engineer" products that are already in the marketplace. Notably the phrase "reverse engineering" is applied to manufactured products that have somehow managed to get around existing patents. In reality it is the manufacturer, rather than the engineer, who has overcome the legal limitiations of the claims of the patent. Unfortunately, for the engineer, the phrase "reverse manufacturing" doesn't doesn't carry quite the same linguistic impact as reverse engineering, the latter term having already passed into common usage.

Patenting is a two-edged sword for inventors and manufacturers. On the one hand a patent provides some protection for novel ideas. On the other hand the public disclosure of the idea, legally imposed on all published patents, provides a rich pool of opportunity for reverse engineering. Henry Ford has reputedly claimed never to have patented his ideas. He was also reputed to have said that " what I save on patent attorneys I place into a fund for defending patent infringements."

P8.8 As an exercise in reverse engineering it is useful to explore the patent claims of the inventions depicted in Figures 8.41 through 8.45. In each case examine the claims and full descriptions of the invention and offer ideas for reverse engineering an alternative invention. Rank-order the inventions depicted in Figures 8.41 through 8.45 in terms of the strength and uncircumventability of their claims.

8.7 Chapter Summary

In this chapter we explore the idea of open ended MaT projects. After examining a specific case example (Project Metaphor), several sample problems are offered as opportunities for develping these types of MaT projects. The chapter provides some elementary insight into patents and reverse engineering.

REFERENCES AND BIBLIOGRAPHY

Abdel-Rahman, T.M. and Shawki, G.S.A. (1989) Model-aided techniques for creative mechatronics designs. In *Proceedings–Frontiers in Education Conference*, pp.222–228.

Aczel, A.D. (1996) *Fermat's Last Theorem: Unlocking the Secret of an Ancient Mathematical Problem*. Viking.

Adams, J.L. (1974) *Conceptual Blockbusting*. San Francisco: Freeman.

Adhesives: Handbook of adhesive technology. Pizzi, A. and Mittal, K.L. eds.(c.1994) New York: Dekker.

Adhesives: Structural adhesives directory and databook. (1996), 1st ed. London: Chapman & Hall.

Adhesives: Wood Adhesives. (1995) Forest Products Society, Madison, WI: Forest Products Society Press.

Agrawal, R., Kinzel, G.L., Srinivasan, R. and Ishii, K. (1993) Engineering constraint management based on an occurrence matrix approach. *Journal of Mechanical Design*, Transactions of the ASME 115(1):103–109.

Akao, Y., Ed. (1990). *Quality Function Deployment*. Cambridge MA: Productivity Press.

Akao, Y. and Mizuno S. (1994) *QFD: The customer-driven approach to quality planning and deployment*. Asian Productivity Organization.

Alexander, C. (1964) *Notes on the Synthesis of Form*. Cambridge, MA: Harvard University Press.

Alexander, C. and Manheim, M.L.(1965) The design of highway interchanges: An example of a general method for analyzing engineering design problems. *Highway Research Record* (83):48–87.

Alexander, K.C. (1991) Artmaking: Bridge to metaphorical thinking. *Arts in Psychotherapy* 18(2):105–111.

Allison, J.E. and Cole, G.S. (1993) Metal-matrix composites in the automotive industry: Opportunities and challenges. *JOM* 45(1):19–24 (Jan.).

Ambrose, S.A. and Amon, C. H. (1997) Systematic design of a first-year mechanical engineering course at Carnegie Mellon University. *Journal of Engineering Education* 86(2):173–181 (April).

Amirkhan, J.H. (1994) Criterion validity of a coping measure. *Journal of Personality Assessment* 62(2):242-261 (April).

Appleby A.J., Ed. (1987) *Fuel cells: trends in research and applications*. Washington: Hemisphere.

Appleby, A.J. and Foulkes, F.R. (1993) *Fuel Cell Handbook*. Malabar, FL: Krieger.

Arce-Villalobos, O.A. (1993) *Fundamentals of the Design of Bamboo Structures*. Eindhoven: Technische Universiteit.

Atkins, P.W. (2003) *Galileo's Finger: The Ten Great Ideas of Science*, New York: Oxford University Press.

Bailey, R.L. (1980) *Solar-Electrics Research and Development*, Ann Arbor, MI: Ann Arbor Science.

Balmer, A.J., Cotter, C.E. and Harnish, M.(1998) *Mousetrap Powered Cars and Boats*. Round Rock, TX: Doc Fizzix.

Belloc, H. (2002) *Cautionary Tales for Children* (with illustrations by Edward Gorey), Orlando, FL: Harcourt.

Besenhard, J.O., Ed. (1999) *Handbook of Battery Materials*. New York: Wiley-VCH.

Biden, E. and Rogers, R. (2002) A tradeshow for products developed in second year mechanical engineering. In *Proceedings–Frontiers in Education Conference*, (2):F2D/23-F2D/26.

Bisantz, A.M., Roth, E., Brickman, B., Gosbee, L.L., Hettinger, L. and McKinney, J. (2003) Integrating cognitive analyses in a large-scale system design process. *International Journal of Human Computer Studies* 58(2):177–206.

Blackowiak, A.D., Rand, R.H. and Kaplan, H. (1997) The dynamics of the Celt with second-order averaging and computer algebra. *Proceedings of ASME Design Engineering Technical Conference*, September 14–17, Sacramento, CA, Paper DETC97/VIB–4103.

Boden, M. (1979) *Jean Piaget*. London: Penguin.

Brown, J.R. (1991) Laboratory of the Mind: Thought Experiments in the Natural Sciences. London: Routledge.

Brown, J.R. (2002) Thought Experiments, In *The Stanford Encyclopedia of Philosophy* (http://plato.stanford.edu/), E.N. Zalta Ed.

Bruning, R. and Horn, C. (2000) Developing motivation to write. *Educational Psychologist, Special issue: Writing development: The role of cognitive, motivational, and social/contextual factors* 35(1):25–37.

Bullock, G.N., Denham, M.J., Parmee, I.C. and Wade, J.G. (1995) Developments in the use of the genetic algorithm in engineering design. *Design Studies* 16(4):507.

Byo, J.L. (1993) The influence of textural and timbral factors on the ability of music majors to detect performance errors. *Journal of Research in Music Education* 41(2):156–167.

Calladine, C.R. (1978) Buckminster Fuller's "tensegrity" structures and Clerk Maxwell's rules for the construction of stiff frames. *International Journal of Solids and Structures* 14(2):161–172.

Calladine, C.R. (1983) *Theory of Shell Structures*. New York: Cambridge University Press.

Campbell, D. (2001) The long and winding (and frequently bumpy) road to successful client engagement: One team's journey. *System Dynamics Review* 17(3):195-215.

Canter, L. (1977) *Environmental Impact Assessment*. New York: McGraw-Hill.

Capp, A., Frazetta, F. and Kitchen, D. (1960-1961) *Al Capp's Li'l Abner: The Frazetta Years : Volume 4*. Oakland CA: Darkhorse.

Carson, D.K. and Runco, M.A. (1999) Creative problem solving and problem finding in young adults: Interconnections with stress, hassles, and coping abilities. *Journal of Creative Behavior* 33(3):167–190.

Chanute, O. (1897) Recent experiments in gliding flight. In James Means, *Aeronautical Annual*, Boston (3):37.

Chen, S.-J. and Lin, L. (2002) A project task coordination model for team organization in concurrent engineering. *Concurrent Engineering Research and Applications* 10(3):187–202.

Chen, X. and Hoffmann, C.M. (1995) On editability of feature-based design. *Computer Aided Design* 27(12):905–914.

Christiansen, A.W. and Conner, A.H. (1996) *Wood Adhesives*. Proceedings of a symposium sponsored by the USDA Forest Service, Forest Products Laboratory, June 29–30, 1995, Portland, OR. Madison, WI: Forest Products Society.

Clancey, W.J. (1997) *Situated Cognition: On Human Knowledge and Computer Representation*. New York: Cambridge University Press.

Cohen, L. (1996) *Quality Function Deployment: How to Make QFD Work for You*, Reading, MA: Addison-Wesley, (Reviewed in the *Journal of Product Innovation Management*, March 1996:183-184).

Composites Bonding – see Damico *et al.*

Connor-Greene, P.A. (2002) Problem-based service learning: The evolution of a team project. *Teaching of Psychology* 29(3):193–197.

Cooper, P.W. and Kurowski, S.R. (1996) *Introduction to the Technology of Explosives*. New York: VCH.

Cottrell, D.S. and Ressler, S.J. (1997) Integrating design projects into an introductory course in mechanics of materials. *ASEE Annual Conference Proceedings*.

Courant, R. and Robbins, H. (1996) *What is Mathematics?*; 2nd. Edition, Revised by Stewart, I. New York: Oxford University Press.

Cox, M.V. and Rowlands, A. (2000) The effect of three different educational approaches on children's drawing ability: Steiner, Montessori and traditional. *British Journal of Educational Psychology* 70(4):485–503.

Coxeter, H.S.M. (1973) *Regular Polytopes*, New York: Dover.

Crick, F. and Watson, J. (1953) A structure for deoxyribose nucleic acid. *Nature* (171):737.

Crompton, T.R. (2000) *Battery Reference Book*. Oxford: Newnes.

Crouch, T. (1981) *A Dream of Wings: Americans and the Airplane*, 1875–1905. New York: W.W. Norton.

Daetz, D., Barnard, W. and Norman, R. (1995) *Customer Integration: The Quality Function Deployment (QFD) Leader's Guide for Decision Making*. New York: Wiley.

Damico, D.J. Wilkins T.L., Jr., and Niks, S.L.F., Eds. (1994) *Composites Bonding*. Philadelphia: ASTM.

Davidson, J. K. and Hunt, K. H. (2004) *Robots and Screw Theory*. New York: Oxford University Press.

de Bono, E. (1985) *Six Thinking Hats*. Boston: Little, Brown, and Co.

de Bono, E. (1978) *Teaching Thinking*. Harmondswort: Penguin.

de Bono, E. (1970) *Lateral thinking: Creativity step by step*. New York: Harper & Row.

Dell, R.M. and Rand, D.A.J. (2001) *Understanding Batteries*. Cambridge: Royal Society of Chemistry.

De Mille, R. (1994) *Put Your Mother on the Ceiling*. New York: Viking.

De Paul, G. and Mercer, J. (1956) *Li'l Abner: The Musical*.

Design Matrix: *http://www.designmatrix.com/tools/cluster_tools.html*.

Designing Plastic Parts for Assembly – see Tres.

De Simone, D.V. (1968) *Education for Innovation*. Oxford: Pergamon.

Dogan, K. and Goetschalckx, M. (1999) Primal decomposition method for the integrated design of multi-period production-distribution systems. *IIE Transactions* (Institute of Industrial Engineers) 31(11):1027–1036.

Dowie, M. (1977) Pinto madness, *Mother Jones* p.35 (September/October).

Du, H. and Nair, S.S. (1996) Design of a train separator for hydraulic capsule pipeline control. *ASME Fluid Power Systems and Technology* 3:65–70.

Duckworth, E., See Piaget.

Durfee, W.K. (1994) Engineering education gets real. *Technology Review* 97(2): 42–51 (Feb.–Mar.).

Dutta, R.D. (1991) A model of change in organizational health to improve quality of life. *Social Science International*, 7(2):6–11 (July).

D'Zurilla, T. and Chang, E.C. (1995) The relations between social problem solving and coping. *Cognitive Therapy & Research* 19(5):547–562 (Oct.).

Eddy, P., Potter, E. and Page, B. (1976) *Destination Disaster*. London: Hart-Davies, McGibbon.

Edison, T. (1932) An interview reported in *Harper's Monthly Magazine*, September.

Egelhoff, C.J. and Odom, E.M. (1999) Machine design: Where the action should be. In *Proceedings - Frontiers in Education Conference* (2): 12c5:13–18.

Einstein, A. and Infeld, L. (1938) *Evolution of Physics*. New York: Simon & Schuster.

Epstein, H.T.(1970) *A Strategy for Education*. New York: Oxford University Press.

Falk, G. (1982) An empirical study measuring conflict in problem-solving groups which are assigned different decision rules. *Human Relations* 35(12):1123–1138.

Field, B.W., Weir, J.G. and Burvill, C.R. (2003) Some misconceptions in student's understanding and use of physics for solving problems in engineering mechanics. *University of Melbourne Design Report DR0203*, University of Melbourne: Desktop Design Centre.

Fodor, E.M. and Carver, R.A. (2000) Achievement and power motives, performance feedback, and creativity. *Journal of Research in Personality* 34(4):380–396 (Dec.).

Fodor, E. M. and Roffe-Steinrotter, D. (1998) Rogerian leadership style and creativity. *Journal of Research in Personality* 32(2):236–242 (June).

French, M. (1988) *Invention and Evolution: Design in Nature and Engineering*. New York: Cambridge University Press.

French, M. (1985) *Conceptual Design for Engineers*. London: The Design Council.

Fuchs, H.O.(1976) Private communications at Stanford University.

Fuchs, H.O. and Steidel, R.F., Jr. (1973) *10 Cases in Engineering Design*. Essex, UK: Longman.

Gedanken (German): thoughts, ideas; see also Brown, J. R.(2002); URL: <*http://plato.stanford.edu/archives/sum2002/entries/thought-experiment/*>.

Gere, J.M. and Timoshenko, S.P. (2002) *Mechanics of Materials*. London: Chapman & Hall.

Gibbs-Smith, C. and Reese, G. (1978) *The Inventions of Leonardo Da Vinci*. Oxford: Phaidon.

Gladman, M. and Lancaster, S. (2003) A Review of the Behaviour Assessment System for Children. *School Psychology International* 24(3):276–291 (Aug.).

Glickauf-Hughes, C. and Campbell, L.F. (1991) Experiential supervision: Applied techniques for a case presentation approach. *Psychotherapy: Theory, Research, Practice, Training* 28(4):625–635

Gordon, J.E. (1979) *Structures or why things don't fall through the floor*. Harmondsworth: Penguin.

Green, D.W., Winandy, J.E. and Kretschmann, D.E. (1999) Mechanical properties of wood. In Wood handbook: Wood as an engineering material. *Forest Products Laboratory General Technical Report FPL; GTR*-113:4.1–4. Madison, WI: USDA Forest Service.

Gruber, H. and Voneche, J., Eds. (1977) *The Essential Piaget*. New York: Basic.

Gryskiewicz, N.D. and Tullar, W.L. (1995) The relationship between personality type and creativity style among managers. *Journal of Psychological Type* (32):30–35.

Güémez, J., Valiente, R., Fiolhais, C. and Fiolhais, M. (2003) Experiments with a sunbird. *American Journal of Physics* (71):1264.

Guinta, L.R. and Praizler, N.C. (1993) *The QFD Book: The Team Approach to Solving Problems and Satisfying Customers through Quality Function Deployment*. New York: AMACOM; Reviewed in the *Journal of Product Innovation Management*, June 1994:275–276.

Handbook of Adhesive Technology – see Pizzi.

Hart-Davies, A. (1999) *The World's Stupidest Inventions*. Reading: Cox and Wyman.

Härtel, H. (1994) COLOS: Conceptual learning of science. In de Jong, T. and Sarti, L., Eds. *Design and Production of Multimedia and Simulation-based Learning*. Norwell, MA: Kluwer pp. 189–217.

Hauser, J.R. and Clausing, D. (1988) The house of quality. *Harvard Business Review* 66(3):67–73.

Hawking, S. W. (1988) *A Brief History of Time*. New York: Bantam.

Henry, J. (1991) *Creative Management*. London: Sage.

Herring, A.M. (1897) Recent advances toward a solution of the problem of the century. *Aeronautical Annual* (3):70–74.

Hewitt, G.F. (2000) *Introduction to Nuclear Power*. New York: Taylor & Francis.

Hibbard, W.J. and Hibbard, R.L. (1995) Generating excitement about mechanical engineering by using hands-on projects. In *ASEE Annual Conference Proceedings* (2, *Investing in the Future*) pp.2471–2476.

Hodge, B.K. and Taylor. R.P. (1999) *Analysis and Design of Energy Systems*. Upper Saddle River, NJ: Prentice Hall.

Janssen J.J.A. (1995) *Building with Bamboo – A Handbook*. London: Intermediate Technology.

Jewkes, J., Sawers, D. and Stillerman, R. (1969) *The Sources of Invention*. New York: W.W. Norton.

Johnson, P.R. (1955) Leaning Tower of Lire. *American Journal of Physics* 23:240.

Johnston, W.D., Jr. (1980) *Solar Voltaic Cells*. New York: Dekker.

Jones, J.B. and Dugan, R.E.(1996) *Engineering Thermodynamics*. Englewood Cliffs, NJ: Prentice Hall.

Jones, R.K. and Wang, E. (2001) Experiences with an engineering technology course for education majors. *Frontiers in Education Conference* 1:T1E/4–T1E/8. (Oct. 10–13).

Jung, C.G. and Hull, R.F.C., Trans. (1990) *The basic writings of Carl Gustav Jung*. Bollingen series, No. 20. Princeton, NJ: Princeton University Press.

Kaisinger, R. (2002) *Build a Better Mousetrap, Make Classic Inventions, Discover Your Problem Solving Genius and Take the Inventor's Challenge*. Hoboken, NJ: Wiley.

Kalamaras, G.S. (1997) Computer-based system for supporting decisions for tunneling in rock under conditions of uncertainty. *International Journal of Rock Mechanics and Mining Sciences* 34(3–4):588.

Kelly, F.C. (1996) *Miracle at Kitty Hawk, The Letters of Wilbur and Orville Wright*. New York: Da Capo.

King, R. (1995) *Designing Products and Services That Customers Want*. Cambridge MA: Productivity Press.

Klein, J.P. (1975) Socratic dialogue vs. behavioural practice in the development of coping skills. *Alberta Journal of Educational Research* 21(4):255–261.

Klenk, P., Barcus, K. and Ybarra, G.A. (2002) Techtronics: Hands-on exploration of technology in everyday life. In *Proceedings – Frontiers in Education Conference* (1):T3C/18-T3C/23.

Kondratieff, N. (1984) *The Long Wave Cycle*. New York: Richardson and Snyder.

Korfmacher, J. and Spicer, P. (2002) Toward an understanding of the child's experience in a Montessori Early Head Start program. *Infant Mental Health Journal, Special Issue: Early Head Start* 23(1–2):197–212, (Feb.).

Ledsome, C. (1987) *Engineering Design Teaching Aids*. London: The Design Council.

Lienhard, J.H., IV and Lienhard, J.H., V (2004) *A Heat Transfer Text Book*. Lexington MA: Phlogiston.

Likert, D, and Likert, J. (1978) A method for coping with conflict in problem-solving groups. *Group & Organization Studies* 3(4):427–434, (Dec.).

Lillard, P.P. (1997) *Montessori in the Classroom: A Teacher's Account of How Children Really Learn*. New York: Random House.

Lillard, P.P. (1996) *Montessori Today: A Comprehensive Approach to Education From Birth to Adulthood*. New York: Random House.

Logan E., Ed. (1995) *Handbook of Turbomachinery*. New York: Dekker.

Logan, E. (1993) *Turbomachinery*. New York: Dekker.

Lotz, N.W. (1995) Trying to coincide the inner- and outer world: The Socratic dialogue. *Communication & Cognition* 28(2-3):165–186.

Love, L.J. and Dickerson, S.L. (1994) Mechanical systems lab: A real life approach to teaching engineering. *ASME, Dynamic Systems and Control Division, Publication DSC* (55–2), *Dynamic Systems and Control* 2:1063–1066.

McKim, R.H. (1978) The Imaginarium: An environment and program for opening the mind's eye. In *Visual Learning, Thinking and Communication*, B.S. Randhawa and W.C. Coffman, Eds. New York: Academic Press.

McKim, R.H.(1972) *Experiences in Visual Thinking*. Monterey, CA: Brooks-Cole.

Martinez-Pons, M. (2001)*The Psychology of Teaching and Learning: A Three-step Approach*. New York: Continuum.

Mayhew, Y. and Hollingsworth, M.(1996) *Engineering Thermodynamics: Work and Heat transfer Solutions Manual*. Harlow: Longman.

Mensch, G. (1975) *Das Technogische Patt* (Stalemate in Technology). Frankfurt-am Main:Umschau Verlag.

Miranda, M.L. (2000) Developmentally appropriate practice in a Yamaha Music School. *Journal of Research in Music Education* 48(4):294–309.

Mizuno, S. and Akao, Y., Eds. (1994) *QFD: The Customer-Driven Approach to Quality Planning and Deployment*. Cambridge MA: Productivity.

Moje, E.B., Collazo, T., Carrillo, R. and Marx R.W. (2001) Maestro, what is 'quality'?: Language, literacy and discourse in project-based science. *Journal of Research in Science Teaching* 38(4):469–498 (April).

Montessori, M. (1976) *Education for Human Development: Understanding Montessori*. New York: Schocken.

Moore, J. (2005) Is higher education ready for transformative learning? – A question explored in the study of sustainability. *Journal of Transformative Education* 3(1):76–91 (Jan.).

Moran, M.J. and Shapiro, H.N. (2000) *Fundamentals of Engineering Thermodynamics*. New York: Wiley.

Motro, R. (2003) *TENSEGRITY: Structural Systems for the Future*. London: Kogan Page.

Murphy, P.S. (1994) The contribution of art therapy to the dissociative disorders. *Art Therapy* 11(1):43–47.

Narri, Y. and Mummadi, V. (1999) Adaptive controller for PV supplied buck-boost converter. In *Proceedings of the International Conference on Power Electronics and Drive Systems* 2:789–793.

Newman, W.R. (2002) *Gehennical Fire: The Lives of George Starkey, an American Alchemist in the Scientific Revolution*. Chikago: University of Chicago Press.

Ng, W.-Y. (1991) Interactive descriptive graphical approach to data analysis for trade-off decisions in multi-objective programming. *Information and Decision Technologies* 17(2):133–149.

Niku, S.B. (1995) Metamorphic mechanical dissection and design in freshman engineering courses. In *ASEE Annual Conference Proceedings, v. 2, Investing in the Future*:2035–2038.

Nolan, D.P. (1996) *Handbook of Fire and Explosion Protection Engineering Principles for Oil, Gas, Chemical, and Related Facilities*. Westwood, NJ: Noyes.

Nuclear Energy Agency (2003) *Nuclear Energy Today*. Paris: Nuclear Energy Agency, Organisation for Economic Co-operation and Development.

Ocean Energy: See also www.eere.energy.gov/RE/ocean.html.

Ord-Hume, A.W.J.G.(1977) *Perpetual motion: The history of an obsession*. New York: St Martin's.

Otsuka, K. and Wayman, C. M., Eds. (1998) *Shape Memory Materials*. Cambridge, New York : Cambridge University Press.

Pahl, G. and Beitz, W. (1996) *Engineering Design: A Systematic Approach*, 2nd ed. trans. Ken Wallace, Lucienne Blessing and Frank Bauert; ed. Ken Wallace. London: Springer.

Papastavridis, J.G.(2002) *Analytical Mechanics: A Comprehensive Treatise on the Dynamics of Constrained Systems, For Engineers, Physicists, and Mathematicians*. New York: Oxford University Press.

Park, R. (2000) *Voodoo Science: The Road from Foolishness to Fraud*. New York: Oxford University Press.

Parmee, I.C. (1997) Evolutionary computing for conceptual and detailed design. In Quagliarella *et al.*, Eds. (1999) *Genetic Algorithms and Computer Science*. New York: Wiley, pp.133–152.

Parrott, C.A. and Strongman, K.T. (1985) Utilization of visual imagery in creative performance. *Journal of Mental Imagery* 9(1):53–66.

Perry, N.E., VandeKamp, K.O., Mercer, L.K. and Nordby, C.J. (2002) Investigating teacher-student interactions that foster self-regulated learning. *Educational Psychologist* 37(1):5–15 (Mar.).

Piaget, J. (1970) *Genetic Epistemology*. trans. E. Duckworth, New York: Norton.

Pizzi, A. and Mittal, K.L., Eds. (2003) *Handbook of Adhesive Technology*. New York: Dekker.

Plato and Guthrie W.K.C. (1956) *Protagoras and Meno*. Harmondsworth: Penguin.

Pocius, A.V. (2002) *Adhesion and Adhesives Technology: An Introduction*. Munich: Hanser.

Popper, K. (1963) *Conjectures and refutations: The growth of scientific knowledge*. New York Routledge.

Radjou, N., Cameron, R., Kinikin, E. and Herbert, L.(2004) *Innovation Networks, A New Market Structure Will Revitalize Invention-To-Innovation Cycles*. Boston: Forrester Research.

Reiner-Decher, R. (1994) *Energy Conversion: Systems, Flow Physics, and Engineering*. New York: Oxford University Press.

Reiter-Palmon, R., Mumford, M.D., O'Connor, B.J. and Runco, M.A. (1997)Problem construction and creativity: The role of ability, cue consistency and active processing. *Creativity Research Journal* 10(1):9–23.

Rittel, H. (1987) Keynote address. In *International Conference on Engineering Design*, Boston: ASME .

Ross, M.E., Salisbury-Glennon, J.D., Guarino, A., Reed, C.J. and Marshall, M. (2003) Situated Self-regulation: Modeling the interrelationships among instruction, assessment, learning strategies and academic performance. *Educational Research & Evaluation* 9(2):189–209 (June).

Samuel, A.E. (1986) Student centered teaching in engineering design. *Instructional Science* 15:47–79.

Samuel A.E. (1984) Educational objectives in engineering design. *Instructional Science* 13: 243–273.

Samuel, A.E. and Lewis W.P. (1986) A Socratic approach to design teaching. In *Proceedings National Conference on Teaching Designers for 21st. Century*. University of NSW, pp.143–151 (Feb.).

Samuel, A.E. and Lewis, W.P. (1974) Student response to an open-ended design project. *I.E. Aust. Mech. Engineering Transactions* MC-10(1):23–27.

Samuel, A.E. and Weir, J.G. (1999) *Introduction to Enginering Design: Modelling, Synthesis and Problem Solving Strategies*. London: Elsevier Butterworth.

Samuel, A.E. and Weir, J.G. (1995) Concept learning in engineering design using cognitive maps. *International Conference on Engineering Design,* ICED 95, Prague, August, 22–24: pp.303–311.

Samuel, A.E., Lewis, W.P. and Weir, J.G. (2000) A common language of design for excellence. In *Proceedings Engineering Design Conference 2000*, Brunel, Uxbridge, July, pp.439–448.

Samuels, M. and Samuels, N. (1975) *Seeing with the Mind's Eye, The History, Techniques and Uses of Visualisation*. New York: Random House.

Seabrook, W. (1941) *Dr. Wood: Modern wizard of the laboratory*. Orlando, FL: Harcourt.

Seffen, K.A. (2001) On the behaviour of folded tape-springs. *Journal of Applied Mechanics, Trans. ASME* (68):369–375.

Setlow, R.B. and Pollard, E.C. (1962) *Molecular Biophysics*. Reading, MA: Addison Wesley.

Shute, N. (1951) *No Highway*. New York: Morrow.

Simon, H.A. (1996) *The Sciences of the Artificial*. 3rd ed. Cambridge, MA: The MIT press.

Simon, H.A. (1978) Rational decision-making in business organizations. *Nobel Memorial Lecture*, 8 December 1978.

Simon, H.A. (1952) A Behavioral model of rational choice. *Quarterly Journal of Economics* 69:99–118.

Simon, H. A. and Kaplan, C.A. (1989) Foundations of cognitive science. In Posner, M.I., Ed. *Foundations of Cognitive Science*, Cambridge MA: MIT Press.

Smith, D.K. and Alexander, R.C. (1988) *Fumbling the Future: How Xerox Invented, Then Ignored, The First Personal Computer*. New York: Morrow.

Snelson, K.D. (2002) *Space Frame Structure Made by 3-D Weaving of Rod Members*. US Patent 2002/0,081,936.

Snelson, K.D. (1973) *Tensegrity Masts*. Bolinas, CA: Shelter.

Sobel, D. (2000) *Galileo's Daughter*. London: Fourth Estate.

Sobel, D. (1995) *Longitude*. Harmodsworth: Penguin.

Sodan, A.C. (1999) Toward successful personal work and relations: Applying a yin/yang model for classification and synthesis. *Social Behavior and Personality* 27(1):39–71.

Srinivasan, A.V. and McFarland, D.M. (2000) *Smart Structures: Analysis and Design*. Cambridge: Cambridge University Press.

Terninko, J., Ed. (1997) *Step by Step QFD: Customer Driven Product Design*. Boca Raton, FL: CRC.

Terzis, G. (2001) How crosstalk creates vision-related eureka moments. *Philosophical Psychology* 14(4):391–421 (Dec.).

Timoshenko, S.P. (1981) *Strength of materials*. Volume 1, Melbourne, FL: Krieger.

Timoshenko, S.P. (1953) *History of strength of materials*. New York: McGraw-Hill.

Tres, P.A. (2003) *Designing Plastic Parts for Assembly*. 5th ed. Munich:Hanser Gardner.

Vaughn, M. and Stairs, A. (2000) Delicate balance: The praxis of empowerment at a midwestern Montessori school. In *National Communication Association Conference*, Seattle, WA.

Vesper, K.H. (1975) *Engineers at work: A Casebook*. Boston:Houghton Mifflin.

Vidal, F. (1994) *Piaget Before Piaget*. Cambridge MA: Harvard University Press.

Walker, J. (1979) The Amateur Scientist, *Scientific American* 241:172–184.

Weber, R.L. (1973) *A Random Walk in Science*. London: The Institute of Physics. p.77., N-Rays, based on Seabrook (1941).

Weilerstein, P.J. (1999) Fostering applied innovation in higher education: The National Collegiate Inventors and Innovators Alliance. *Frontiers in Education Conference* 1:11a6-1–11a6-3.

Wentzel, K.R and Watkins, D.E. (2002) Peer relationships and collaborative learning as contexts for academic enablers. *School Psychology Review* 31(3):366-377.

Werner, J.M and Lester, S.W. (2001) Applying a team effectiveness framework to the performance of student case teams. *Human Resource Development Quarterly* 12(4):385-402.

Wesenberg, P. (1994) Bridging the individual-social divide: A new perspective for understanding and stimulating creativity in organizations. *Journal of Creative Behavior* 28(3):177–192.

White, B. (1983) Sources of difficulty in understanding Newtonian dynamics. *Cognitive Science* 7(1):41–65A.

White, B.Y. and Frederiksen, J.R. (1998) Inquiry, modeling, and metacognition: Making science accessible to all students. *Cognition and Instruction* 16(1):3–118.

Wiles, A. (1995) Modular elliptic curves and Fermat's Last Theorem. *Annals of Mathematics* 142:443–551.

Williams, R. (2002) *Retooling: A historian confronts technological change.* Cambridge: The MIT Press.

Winchester, S. (1999) *The surgeon of Crowthorne.* London: Penguin.

Wohl, B. (1979) *The China Syndrome.* New York: Bantam Press.

Wood Adhesives (1996)– see Christiansen and Conner.

Wood handbook—Wood as an engineering material. Forest Products Laboratory. 1999. Gen. Tech. Rep. FPL-GTR-113. Madison, WI: U.S. Department of Agriculture, Forest Service, Forest Products Laboratory.

Xanthanite Explosives (1999) *Performance Parameters of Condensed High Explosives,* Enfield, S.A.: Xanthanite Explosives Publications.

Yamaha music : www.yamaha-music.com.sg/main_aboutus.html.

Yoji, A., Ed. (1990) *Quality Function Deployment: Integrating Customer Requirements into Product Design.* Cambridge, MA: Productivity Press.

Zeigler, T.R. (1987) *Portable Shelter Assemblies.* US Patent 4,689,932.

Zeigler, T.R. (1997) *Family of Collapsible Structures & a Method of Making a Family of Collapsible Structures.* US Patent 5,651,22.

Zucker, R.D. and Biblarz, O. (2002) *Fundamentals of Gas Dynamics.* Hoboken, NJ: Wiley.

Engineering Case Library

Brewerton, F.J.(1970) The engineering case study–An interdisciplinary comparison. *SAE Paper 70003*.

Buck, D. and Stucki, D.J. (2001) JKarelRobot: a case study in supporting levels of cognitive development in the computer science curriculum. In *Proceedings of the thirty second SIGCSE technical symposium on Computer Science Education* pp.16–20.

Delatte, N.J., Jr. (1997) Failure case studies and ethics in engineering mechanics courses. *Journal of Professional Issues in Engineering Education and Practice* 123(3):111–116.

Dunn-Rankin, D., Bobrow, J.E., Mease, K.D. and McCarthy, J.M. (1998) Engineering design in industry:Teaching students and faculty to apply engineering science in design, *Journal of Engineering Education* 87(3):219–222.

Ellis, C.D., Rakowski, R.T. and Marsh, P.T.C. (1985) Design project case study: Capacitor design for high temperature applications. *International Journal of Applied Engineering Education* 1(5):349–353.

Feldy, E.C., Ed. (2003) Mechanical engineering design education-2001: Issues and case studies. *ASME Design Engineering Division* (113):183 pp.

Henderson, J.M. and Steidel, R.F., Jr. (1974) Engineering case studies — The academic/professional link. *IEE Conference Publication* 11:342–346.

Hirsch, R.A. (1974) The case method as a means of achieving industry-university interaction in design education. *ASME Paper 74-WA/DE-27*.

Kardos, G.(1992) Getting the most from make-and-break projects. *Innovation, Teaching and Management* 205.

Kardos, G. (1974) Engineering cases—Feedback from industry. *ASME Paper 74-WA/DE-31*.

Kardos, G. and Smith, C.O. (1983) Engineering cases as tools for teaching design. *Mechanical Engineering* 105(3):68–71.

Lichy, W.H. and Mariotti, J.J. (1970) The use of case methods at General Motors Institute. *SAE Preprint 70005*.

McKechnie, R.E. (1975) The engineering case study as an aid to career advancement. *ASME Paper 75-DE-19*.

McNair, M.P. (1954) *The Case Method At The Harvard Business School*. New York:McGraw-Hill.

Midha, A., Turcic, D.A. and Bosnik, J.R.(1981) Creativity in the classroom – a collection of case studies in linkage synthesis. *Mechanism & Machine Theory* 19(1):25–44.

Newstetter, W. C. (1998) Of green monkeys and failed affordances: A case study of a mechanical engineering design course. *Research in Engineering Design – Theory, Applications, and Concurrent Engineering* 10(2):118–128.

Noymer, P.D., Hazen, M.U. and Yao, S.C. (1998) Integrated thermal science course for third-year mechanical engineering students. *ASME Heat Transfer Division* 361-3:49–53.

Samuel, A.E. and Lewis, W.P. (1987) Teaching less and learning more – a Socratic approach to engineering design education. *ASME Transactions* 1055–1065.

Seif, M.A. (1997) Design case studies: A practical approach for teaching machine design. *Frontiers in Education Conference* 3:1579–1582.

Smith, C.O. and Kardos, G. (1987) Processes for teaching design processes. *ASME transactions* 1081–1088.

Vesper K. H. and Adams J.L. (1971) Teaching objectives, style, and effect with the Case Method. *Engineering Education* 61(7):831–833

Readings for MaT Project Planning

Aaseng, N. (2000) *Construction: Building the impossible*. Minneapolis: Oliver.

Anderson, N. (1983) *Ferris Wheels*. New York: Pantheon.

Balmer, A.J. (1998) *Mousetrap Cars:The Secrets to Success*. Round Rock, TX: Doc Fizzix.

Bochinski, J.B.(2004) *The Complete Workbook for Science Fair Projects*. New York:Wiley.

Bochinski, J.B.(1996) *The Complete Handbook of Science Fair Projects*. Hoboken, NJ: Wiley.

Bransford, B.D. and Stein, B.S. (1984) *The IDEAL Problem Solver*. New York: W. H. Freeman.

Duffy, T. (2000) *The Clock*. New York: Atheneum.

French, M. (1992) *Form, Structure and Mechanism*. New York: McMillan.

APPENDIX A1
A PRIMER ON MECHANICS

When we were carrying on our wind-tunnel work we had no thought of ever trying to build a power aeroplane. We did that work just for the fun we got out of learning new truths.
Orville Wright, 1928 (in Kelly, 1996)

In the early twentieth century Albert Einstein revolutionised theoretical physics by overturning Newton's view of gravity. With it went Newton's view of the universe as permanent and unchanging. Although Einstein himself resisted it, the equations seemed to predict a dynamic universe.
Stephen Hawking, 1999

Mechanics is the science of forces acting on bodies at rest and on bodies in motion. The whole of mechanics is subdivided into statics, kinematics and dynamics.

- Statics deals with forces on bodies at rest and in general addresses the equilibrium of these forces by both external and internal reactions.

- Kinematics is the study of the relative motion of one or more connected or disconnected bodies or particles, without any reference to forces acting on them.

- Dynamics is the study of forces acting on bodies in motion.

A1.1 STATICS

A1.1.1 FORCES

As a helpful introduction to this aspect of mechanics we examine the types of forces and combinations of forces that can act on bodies or systems of bodies. We often talk about forces, but we equally often overlook the fact that a force is not an easily measured quantity. In fact in a most general way the concept (idea) of a force is based entirely on the "effect" it can create.

When we weigh ourselves on a simple bathroom scale, the effect created by our mass acted on by gravity is to displace a spring inside the scale. A pointer attached to the spring will then indicate the "force" of gravity acting on our body mass.

Figure A1. Schematic view of weightlifter action   Figure A2. Weightlifter in action

243

Similarly, when a weightlifter lifts a weight (refer to Figures A1 and A2), his muscles and bones react to the force of gravity acting on the mass of the weight being lifted. Isaac Newton recognised that two masses attract each other in proportion to their masses and inversely proportional to the square of the distance between their respective centres. The constant of proportionality in this relation is the "gravitational constant" G.

Acceleration due to gravity $g = (G \times M)/r^2$, where

G = the gravitational constant = 6.673×10^{-11} m3 kg^{-1} s^{-2}

M = mass of the earth = 5.974×10^{24} kg

R = the radius of the earth = 6,378,140 m at the equator.

The earth is a flattened sphere (referred to by geometers as a "prolate spheroid") and its radius at the poles is about 2100 m less than at the equator.

The current holder of the world super-heavy class (greater than 105 kg body weight) championship for "clean and jerk"[A1] is Hossein Reza Zadeh of Iran, who lifted 263.5 kg at the Athens Olympics in 2004. The clean and jerk lifting style is shown in the photo of Figure A2. It is instructive to calculate the relative effect on a record such as Reza Zadeh's if the weightlifting competition were held at Quito (capital city of Ecuador, through which the equator passes) and at Svalbard, on the Arctic island of Spitsbergen (at 78 degrees of latitude).

Forces can act as individuals or in pairs forming couples. A force is a vector quantity having magnitude, direction and sense and it is usually depicted as a directed arrow. They can act in a plane (called planar system of forces) and in three dimensions (called a "general system of forces").

A pure couple is a planar system of two equal and opposite forces separated by some distance. The effect of a couple is called a moment and the moment of a couple is measured by the product of the force and the distance between them. Moments are also vector quantities and may be represented by a directed arrow normal to the plane in which the couple acts.

Figure A3 shows a planar system of forces acting on a mathematically thin planar body called a "lamina". The forces are shown acting normal to the lamina at their

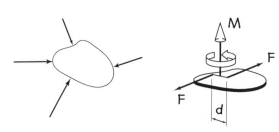

Figure A3 System of forces acting on a lamina

Figure A4 Moment of a couple $M = d \times F$

Figure A5 General system of forces acting on a body

A1 Clean and jerk is a two stage-lifting technique used in weightlifting, with each lifting stage raising the bar about half of the total lift.

point of application. This avoids the introduction of forces acting tangential to the lamina (friction forces, for example). Figure A4 shows a couple acting on a lamina. The moment of this couple is a vector M as shown. The sense of M is positive when pointing out of the lamina as shown. Figure A5 shows a general system of forces acting on a body. The combination of these forces results in a single general mechanics entity called a wrench, which is a combination of a general force and a moment acting on a single line called a screw. The instantaneous action of this wrench on the body is to twist it in a helicoidal field of motion about the screw. Although it is always useful to think of general systems of forces as wrenches, the treatment of screw geometry is beyond the scope of this simple primer.[A2] Moreover, for the purposes of simple structural and dynamic analysis, almost all of the systems of forces acting can be either represented by planar forces and moments, or resolved into such systems of forces and moments.

AI.I.2 PLANAR PINNED STRUCTURES

For systems of planar forces a basic rule of Newtonian mechanics is that for these forces to be in equilibrium (*i.e.*, the lamina should not move under the action of these forces) is that:

(a) Either the lines of actions of all the forces should intersect at one point (*i.e.*, they should be concurrent), or

(b) Their lines of action should be parallel (i.e. their point of concurrency is at infinity according to Euclidian geometry), and

(c) To every action (applied force or moment of forces) there is an equal and opposite reaction (equilibrating force or moment of forces).

When we build simple structures the most important consideration is that the structure should support the loads (forces and moments) imposed on it. These forces and moments will generate internal forces and moments in the material of the structure and eventually these internal forces and moments will cause the material of the structure to break or deflect too much.

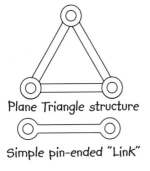

Plane Triangle structure

Simple pin-ended "Link"

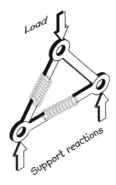

Figure A6 Simple structural elements   Figure A7 Pinned planar structure under load

A2 See Papastravidis (2002) Davidson and K. H. Hunt (2004)

Simple load bearing structures are often constructed from several simple parts connected together. The simplest part of a structure is sometimes referred to as a link, signifying the linking together of several parts of the structure. When such simple structures are subjected to external loads, the loads are distributed into the several links making up the structure. We make various simplifications about the nature and behaviour of such links in structures. These simplifications permit easy evaluation of the forces generated in the links.

Figure A6 shows a simple pin-ended link and the simplest type of plane structure we can construct using only the links and pins inserted at the joints between each pair of links. The following are the simplifications usually applied to allow simple analysis for internal link forces:

- Although the real structure must have some thickness (in the direction perpendicular to the paper on which it is sketched), the applied loads and internal link forces generated will all act in the same plane;

- The pins at the joints and the faces of the links are all "frictionless".

A useful way to think about the forces in the links of such a structure is to imagine the links replaced by springs and also imagine how these springs will deflect under some externally applied load. Figure A7 shows two links replaced by springs and clearly the horizontal link spring will be expanding and the two angled links will be contracting under the applied external loads. These deformations then give us clues about the nature of the internal loads generated in the links by the externally applied loads. In more complicated structures this simple approach will not be easy to apply, but with experience even some complicated three-dimensional structures will yield to this imaginary evaluation. Many commonly used planar structures are built up from such triangles and the resulting trusses are referred to as triangulated trusses. Figure A8 shows some simple triangulated trusses used in bridges and roof construction. The types of forces generated in planar frictionless pin-jointed structures may be classified as follows:

(a) Tensile forces - The structural element that carries only this type of load is called a "tie";

(b) Compressive forces - The structural element that carries only this type of load is called a "strut".

The load-bearing capacity of any member of a structure depends on a combined relationship among applied load, cross-sectional area of the structural element

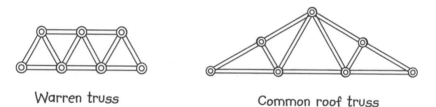

Warren truss   Common roof truss

Figure A8 Simple triangulated planar trusses

exposed to the load and the mechanical properties of the material from which the component is made. For a useful and instructive introduction to mechanical properties of materials interested and enthusiastic readers are referred to J.E. Gordon's *Structures or Why Things Don't Fall Through the Floor*.

The material's properties are usually established by standardised tests involving standard test specimens made from the material of interest. These test specimens have a well-defined area of cross section over which the load is presumed to be acting uniformly. The tests are performed both in tension and compression and, in general for metals, the values obtained for mechanical properties are similar in both types of loading. However, the ultimate failure of a test specimen (or indeed a structural component) in tension is associated with the specimen (or component) physically breaking apart. For the sake of simplicity only the tensile properties are described here.

All elastic materials have the property of elasticity, or the capability of regaining their unloaded shape once the load is removed. The limit of elasticity is the condition when the load applied is so great that the material can no longer resume its former shape once the load is removed. In metals this limiting condition results in some permanent deformation of the material and the material is said to have sustained permanent set or to have sustained plastic deformation. The reference to plastic signifies the distinctive difference between elastic materials and plastic materials, the latter being unable to regain shape after a load is removed. Figure A9 is a graphic representation of the load deformation behaviour for a ductile metal.

The load and deformation are made universally applicable for the specific material by expressing them respectively as stress (load/unit area) and strain (deformation/unit length). The symbols commonly adopted for these measures are as shown, σ for stress and ε for strain. The ratio of stress/strain in the elastic region of the material, denoted E, is called the "modulus of elasticity", or "Young's modulus", after Thomas Young (1773-1829), the English physicist who first formalised the measurement of this property.

The specific mechanical properties of materials of interest to us are:

- Ultimate tensile strength, defined as the load per unit area of the test specimen (or stress-with the units of pressure) when the specimen breaks. The symbol used for this property is σ_u;

- Elastic tensile strength or more commonly "yield strength" is the load per unit area of the test specimen when the material reaches its elastic limit and begins to deform plastically. The symbol used for this property is σ_y;

- Modulus of elasticity E.

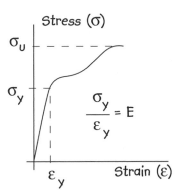

Figure A9 Load-deflection behaviour of a ductile metal (steel, for example)

The relationship among these properties in the elastic region of the material follows the well-

Known form for elastic materials originally expressed by Robert Hooke(1635-1703), now known as Hooke's Law,

$$\sigma = E \times \varepsilon.$$

The task of the designer is to ensure that the loads carried by members in structures do not exceed their elastic limits. In this way we can guarantee that the structure will suffer no plastic deformation. Once the loads acting on the ties and struts in a plane frame are found, the size of the component will be determined by its material properties. When we design for strength (stress-limited design), the area of the strut or tie is found from the ratio of internal load and allowable material stress. For example, a tie constructed from 1.86 mm dry spaghetti will carry a load of approximately 54 N or 5.5 kg mass[A3].

Occasionally we are required to design a structure for minimum deflection (deflection limited design). In that case the elastic properties of the material will govern the size of the tie or strut. In this case we make use of the stress/strain relationship of the material and its elastic modulus as expressed by Hooke's Law for the material.

A further complication is encountered with struts that are slender. A formal definition of the slenderness of struts is derived from elastic theory due to the eighteenth century Swiss mathematician, Leonhardt Euler. Without proof it is stated that a pin-ended strut is regarded as "slender" when

$$L > \sqrt{\frac{2\pi^2 E}{\sigma_y}} \,,$$

where L is the length of the strut, A is its section area, I is called the "second moment of area" of the section and is a measure of the capacity of the section to offer resistance to bending, E is the elastic (Young's) modulus and σ_y is the yielding tensile strength of the material of the tie. Yielding is a form of material failure corresponding to the stress level where the material begins to deform plastically (can no longer recover its original shape when the load is removed).

Slender struts can suffer a form of elastic failure called "buckling". This is a failure form that occurs at much lower loads than the compressive failure load of the material and it is elastic because once the load is removed the strut can recovers its unloaded shape. Unfortunately the effect of a buckled strut is that once it has buckled, it can no longer support any loads and the compressive load originally carried by this strut will be redistributed to other members in the structure. Again without proof, the buckling load in a pin-ended slender strut was derived by Euler as

$$P_{buckling} = \pi^2 \, EI/L^2,$$

where the symbols have the same meaning as before.

Al.l.3 BENDING

When a load is applied transversely to a single structural element, the resulting deformation is referred to as bending. The structural element carrying a bending load only is referred to as a "beam". The distinctive character of beams and their load-

---

A3 Refer to Table 4.6 for data on the strength of dry spaghetti.

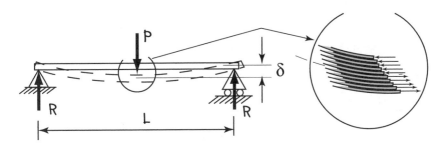

Figure A10 Beam in bending and a grossly exaggerated midsection of the bent beam

carrying behaviour is often described by the type of supports used for reacting against the applied load.

Beam theory applies to small (δ much less than L) deflections only. Under these conditions the curvature of the deflected beam is also small. Another approximation invented to permit evaluation of the internal stresses due to bending is indicated in the exaggerated midsection of the deflected beam in Figure A10. The beam is considered as several thin laminations of material deforming as indicated. The approximation implies that the laminations of the beam behave as do (say) a deck of cards or a stack of thin cardboard when bent.

If we perform an experiment holding a stack of thin cardboard between thumb and forefinger and bending the stack, we find that we need to apply some inward push on the top of the stack and some outward pull on the bottom of the stack. This effect indicates that to permit bending such a stack the upper regions of the cardboard laminations need to experience some compression and the bottom regions must experience some stretching. Somewhere at the centre of the stack there will be no deformation. This central undeformed mathematically thin lamination of a beam is called the neutral surface of the beam. During this bending process the deformed cardboard laminations slide over one another to allow the stretching and compressing to occur.

Figure A11 shows a simple experiment performed to illustrate the model behaviour of laminations in a material when experiencing a small bending load. In Figure A11 (a) the top layers of the cardboard stack are buckled by the compression loads experienced there under bending. Because the thin cardboard laminations are much stronger in tension than in compression, the bottom laminations have slid through the

(a)   (b)

Figure A11 A stack of cardboard held firmly together by spring clips is bent and then straightened

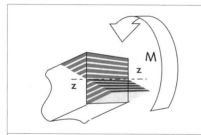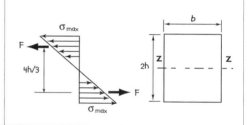

Figure A12 Beam cross section and moment due to bending generating tensile and compressive loads in the beam material	Figure A13 Beam cross section and forces acting due to bending load represented schematically

spring clips holding the stack together. In the solid material beam these laminations will actually experience stretching. This is illustrated in Figure A11 (b) by the extra length gained by the bottom laminations once the cardboard stack is straightened out.   With these approximations we can evaluate the maximum stress in the beam material, generated by the bending load applied.

Figure A12 shows schematically the tensile and compressive loads generated in the beam material by the moment M applied to the beam. Figure A13 shows these tensile and compressive loads expressed as stresses acting on the beam material.

$$F \quad = b \times h \times (\sigma_{max}/2),$$

$$M \quad = 2bh^2(\sigma_{max})/3, \text{ and hence}$$

$$\sigma_{max} = 3M/(2bh^2).$$

For the section shown in Figure A12, the expression $2bh^2/3$, often denoted "Z" in beam theory literature, is called the "modulus of the section". Another important expression ($h \times Z$ or $2bh^3/3$ for the section shown in Figure A13) denoted I_{zz}, is called the "second moment of area" about the axis zz and it is a geometric measure of the section's capacity to resist moments applied to the section about the axis zz.

As noted earlier, beams or structural members subject to transverse loads are often described by the nature of their supports as well as the applied loads. This description helps us to visualise the character of the stresses generated in the beam by the applied loads. The example of Figure A10 above shows a simply supported beam under the action of a single concentrated load. The terminology "simply supported" means that there are no forces or moments imposed on the beam by the supports themselves. The maximum moment in the beam and the associated maximum material stress may be calculated from the equilibrium conditions of the beam and applied load. In the example of Figure A10 the maximum moment M_{max} for a centrally loaded beam is PL/4, and then the resulting maximum tensile or compressive stress in the beam will be at its top and bottom surfaces as indicated in Figure A13.

$$\sigma_{max} = \pm 3 \ PL /(8bh^2),$$

and, by convention, we associate the negative sign with compressive stresses.

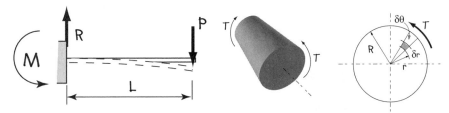

Figure A14 Cantilever beam　　　　　　　Figure A15 Cylindrical rod in torsion

Figure A14 shows a cantilever beam also under the action of a single point load. With this type of beam the one support is built in and provides the reaction load as well as the equilibrating moment necessary to maintain the whole system in equilibrium. Loads on beams may be distributed over the whole beam and in many practical cases of beams both ends may be built into their supports.

The deflection of beams is derived from the differential equation for the shallow curvature of the beam for small deflections. Although deflection theory is outside the scope of this brief introduction, it is worth noting that for the beam examples in Figures A10 and A14 the maximum deflections are $PL^3/48(EI_{zz})$ and $PL^3/3(EI_{zz})$, respectively.[A4]

A1.1.4 TORSION

Bending of a beam resulted in some internal stresses generated in the beam material. We have evaluated these direct stresses in the previous section. However, as the idealised thin laminations of the beam material are stretched and compressed during bending of the beam another stress is also in action. Perhaps we could think of these other types of stresses, called shear stresses, as resulting from the frictional effects of the thin laminations stretching and compressing relative to each other during bending. Most commonly, shear stresses result from a twisting or torsional load applied to a component in a structure.

Figure A15 shows a cylindrical rod under the action of an applied twisting load called Torque T. This torque generates shearing stress in the rod material and Figure A15 is an idealised schematic of how the torque is related to the shear stress in the rod. We can think of the shear stress, usually denoted as "τ", varying from zero at the centre of the rod to τ_{max}, a maximum at the outer surface.

The small component of torque δT acting on an elemental area (shown shaded in Figure A15) is given by

$$\delta T = r \times \tau_r \times \delta r \times r\delta\theta .$$

Because the shear stress varies linearly, from zero in the centre to τ_{max} at the outer surface, we can write the value of the local shear stress at the elemental shaded area as

$$\tau_r = \tau_{max} \; r/R,$$

and the total torque as

$$T = \frac{\tau_{max}}{R} \int_0^R \int_0^{2\pi} r^3 d\theta . dr$$

Integrating and transposing results in

$$\tau_{max} = 2T/\pi R^3.$$

The term $\pi R^4/2$ is a very important geometric property of the section of the cylindrical rod carrying the torque loading, known as the polar moment of area, usually denoted "J" or sometimes "Ip". This property is a direct measure of the capacity of the section to carry the torque load. The term J/R is the polar modulus of section analogous to the modulus $Z = I_{zz}/h$ for the beam in bending.

AI.2 DYNAMICS

AI.2.1 FORCE AND ACCELERATION

When a body, capable of motion, is acted on by a force, in general it will experience acceleration proportional to the force acting on it. If the force and the direction of motion are aligned then by the rule of motion due to Newton

$$F = m\,a,$$

where F is the force acting, m is the mass of the body and a the resulting acceleration of the body. In dealing with problems of motion and acceleration we often make simplifications about level and smooth (frictionless) surfaces, on which our accelerating body moves. Figures A16 and A17 show, respectively, examples of a body accelerating under the action of a force (linear acceleration) and a rotating body under the action of a torque (angular acceleration).

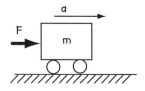

Figure A16 A body accelerating under the action of a force F

Figure A17 A rotating body accelerating under the action of a torque T

Figure A18 A Rumsey engine. Note the concentration of mass at the outer periphery of the flywheel

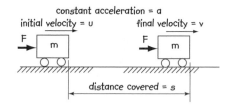

Figure A19 A constant force acting on a body experiencing constant acceleration

For the angular acceleration case, shown in Figure A17, we can apply the rule for rotating bodies (also due to Newton), analogously to linear acceleration

$$T = J\,\alpha\,,$$

where T is the applied torque, α is the angular acceleration and J is a property of the rotating body called the moment of inertia, though sometimes mass moment of inertia to distinguish it from the geometric property we have identified here as the second moment of area, which is occasionally still referred to (quite inappropriately) as the moment of inertia for the section.

The value of this property J depends on the way in which the mass is distributed in the rotating body. Masses farther away from the axis of rotation contribute more to the value of J than those nearer the axis of rotation. This is the reason why flywheels, devices that are used to smooth out the speed variations in intermittently energised devices (such as internal combustion engines) often have their masses concentrated near the outer periphery of the wheel (see Figure A18, a type of engine manufactured by the L.M. Rumsey Manufacturing Co. in St. Louis, Missouri between 1881 and 1917).

There are several useful relationships between time (t), acceleration (a), distance covered (s) and velocity (v) that can be derived using Newton's rules of motion. We only derive these relationships for constant acceleration (or constant force) acting on bodies moving on smooth surfaces without loss. Refer to Figure A19.

For the example shown in Figure A19 we can write that

$$s = \int_0^t v\,dt = \left[\int_0^t a\bullet dt = u + at\right]dt = ut + at^2/2 \ .$$

This equation relating distance covered to elapsed time and acceleration invokes the facts that the body has some initial velocity u and that acceleration is constant in time.

From the work done by the force F we find that the energy change incurred by our moving mass is $M\,(v^2 - u^2)/2 = F.s = M.a.s$. Cancelling the mass and rearranging we get

$$v^2 = u^2 + 2.a.s \ .$$

Similar expressions may be written for constant torque T, acting on a rotating body experiencing constant angular acceleration α . The displacement for rotary motion is expressed as the angle θ rotated. Using arguments similar to those for linear motion above, and using ω_1 as the initial angular velocity and ω_2 as the final angular velocity we get

$$\omega_2 = \omega_1 + \alpha t,$$

$$\theta = \omega_1 t + \omega_2 t^2/2,$$

$$\omega_2^2 = \omega_1^2 + 2.\alpha.\theta.$$

A5 An instructive and easy read on this topic is Peter Atkins' book *Galileo's Finger* (2003). This book, among other interesting material, offers a fine introduction to physical science and the mechanics of work and energy.

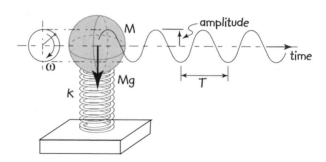

Figure A20   Simple harmonic oscillations of a spring-mass system

AI.2.2 WORK AND ENERGY[A5]

When a force moves a body it is said to do work. Energy of a system (same units as work) is its capacity to perform work. Various forms of energy are interchangeable. For example, thermal energy may be used to do mechanical work (say) on a spring and thereby store potential energy in the spring. Alternatively a falling mass may exchange its potential energy for kinetic energy of motion. Invariably energy is used to perform some form of work and in so doing, as in the exchange of energy from one form to another, we incur some losses. These ideas are formalised in the various rules associated with thermodynamics (the study of heat work and energy exchange). In this brief primer only some simple examples of energy and work are noted, because the study of thermodynamic principles is outside the scope and intention of this appendix.

Probably the simplest form of energy exchange takes place in an oscillating spring, as indicated in Figure A20. The ball of mass M is given a small downward displacement "y" say and then released. The restoring force of the spring will be $k.y$ (Hooke's Law again), where k is the spring stiffness, measured as a force required for unit displacement of the spring. In a sense k is the elastic modulus (see the stress/strain relationship in Figure A8) of the spring. The resulting motion is an oscillation of the spring mass combination about the resting position of the mass (i.e., the mass is displaced Mg/k downward from its position when placed on the uncompressed spring), and this is the position about which the oscillations will occur. The form of the motion is described as simple harmonic motion and corresponds to the projected motion of a point mass executing steady rotation about the rest position of the centre of the mass M. The motion is called harmonic because displacement is a sinusoidal function of time and is related to the way musical instruments vibrate to produce sound. The sound produced by musical instruments is a complex combination of many harmonics.

The differential equation of the motion is written (from Newton's rule $F = Ma$) as

$$Mg - ky = M \, d^2y/dt^2.$$

The solution to this equation is

$$y = A \, \sin(\omega t - \phi) + B.$$

Differentiating this solution twice and substituting into the original equation yields

$$\omega = \sqrt{\frac{k}{M}},$$

and $\beta = k/Mg =$ the original displacement of the spring by M when at rest.

The frequency of oscillation

$$f = \omega/2\pi$$

and the period

$$T = 1/f.$$

The amplitude of the oscillation is found by substituting into these equations the initial conditions.

Clearly, when the mass M is at either extreme of its motion, the spring will have potential energy stored in it. This potential energy is then exchanged with the mass to accelerate it towards the centre of motion where the mass will have maximum kinetic energy. In this simple example we have neglected the mass of the spring. It is an instructive exercise (left for the enthusiastic reader) to consider the influence of the spring mass on the resulting motion.

The value f is a property referred to as the natural frequency or fundamental frequency of spring-mass systems. The form of the relationship

$$f = \frac{1}{2\pi}\sqrt{\frac{k}{M}}$$

may be applied in a wide variety of mechanical systems where we can identify the stiffness k and the vibrating mass M. As a simple example we can apply this fundamental concept to the vibration of the beam of Figure A10. From the deflection formula given $\delta = PL^3/(48EIzz)$ we find that

$$k = P/\delta = 48\ EIzz/L^3.$$

Consequently for that beam carrying a load of P and having a self mass M we can write for the natural frequency of lateral vibration

As an instructive exercise (left for the enthusiastic readers) show that the natural

$$f = \frac{1}{2\pi}\sqrt{\frac{48EI_{zz}}{(M+P/g)L^3}}.$$

frequency of lateral vibration is approximately 61 Hz for a simply supported steel beam with the following properties: L = 1 m, 0.05 m x 0.05 m cross section, carrying a load of 100 N (Steel has a density of 7840 kg/m3).

Another type of system that exhibits harmonic oscillations is a simple pendulum. When the mass of the pendulum is released from some height, its potential energy is converted to kinetic energy of motion and in turn this is reconverted into potential energy as the pendulum swings back to its starting point.

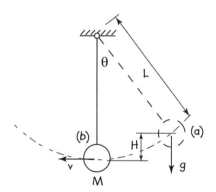

Figure A2I Simple pendulum

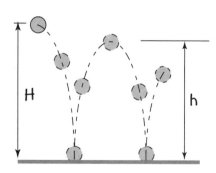

Figure A22 Bouncing ball

Figure A 2I shows a simple pendulum executing small angle oscillations where θ is of the order of 5°. The need for this simplification will become clear when we carry out the simplified analysis of motion for this pendulum.

At location (a) the pendulum mass has potential energy of MgH. At location (b) the mass has lost its potential energy, but gained kinetic energy of $Mv^2/2$.

If the pendulum suffers no losses, then we can equate the two energies and find that

$$gH = v^2/2 \text{ (the mass terms cancelling).}$$

This energy exchange continues indefinitely if the pendulum experiences no losses. The differential equation for this case is found by applying Newton's rule for circular motion, Torque $T = J\, d^2\theta/dt^2$, where J in the case of the simple pendulum shown in Figure A2I is ML^2, because the mass M is assumed concentrated at its centre.

Torque $T = L.\, M.g.\, \sin\theta = M.\, L^2.\, d^2\theta/dt^2$, and simplifying this we get

$$g/L\ \theta - d^2\theta/dt^2 = 0,$$

where we have substituted θ for Sin θ when θ is sufficiently small (as indicated earlier).

This differential equation is of the same form as that found for the spring-mass system above and the pendulum will perform simple harmonic oscillations for small values of θ. The period of oscillation is T and depends only on the length L.

$$T = 2\pi\sqrt{L/g}$$

This is why the simple pendulum is such a fine timekeeper. Considerable effort was devoted by early clockmakers to maintaining L constant, independent of weather conditions.[A6] As a result a commonly used metal for pendulum arms in clocks is invar (short for invariant), a nickel–iron alloy that has been named for its tolerance to modest temperature changes.

Figure A22 shows schematically the behaviour of a bouncing ball dropped onto a hard surface with some initial horizontal velocity. The soft rubber ball, as it falls towards the hard surface, will exchange some of its potential energy for kinetic ener-

---

A6 See, for example, Sobel (1995), the story of James Harrison, the inventor of the marine chronometer.

gy and at impact into yet another form of energy, stored elastic energy. The whole process will reverse itself with some loss in both elastic recovery and in air resistance. The loss in elastic recovery is measured by the "coefficient of restitution" (rebound height/drop height) a property of the elastic material. For the case example shown in Figure A22 the coefficient of restitution is h/H. There are some materials with coefficients of restitution close to unity and the ball in those cases will return to almost the original height from where it was released.

Some typical coefficients of restitution are:

- Basketball   = 0.66,

- Baseball   = 0.54,

- Golf ball   = 0.75.

Some simple expressions for energy are:

- The potential energy of a mass M at height H = M x g x H, where g is the acceleration due to gravity;

- Kinetic energy of a mass m moving with velocity v = m x v^2/2;

- Kinetic energy of a flywheel rotating at angular velocity ω with mass moment of inertia J is $J\omega^2$/2;

- Potential energy of a spring of stiffness k Newton/meter compressed (or deflected) by L meters = k x L^2/2;

- Thermal energy change in a mass M of liquid when this mass experiences a temperature change of ΔT = M x s x ΔT, where s is called the specific heat and is a property of the liquid.

Energy available from burning fuel materials is based on the energy content of the specific fuel and the rate of burning. For example, paraffin (the basic fuel content of candle wax) has 42 kJ/gm of specific fuel energy. As an instructive exercise we could estimate the power (rate of working or rate of dissipating energy) available from a candle. The exercise requires that we burn a candle and time the rate of loss of candle material over some period of time.

A1.3 VERY ELEMENTARY GAS DYNAMICS[A7]

Inasmuch as a compressed gas may be used as a primary energy source in MaT project construction, we need to define and explore some basic ideas about compressed gas behaviour.

Almost all the basic behaviour of gases was developed by scientists in the eighteenth and nieteenth centuries, when performing experiments that involved heating and pressurizing gases. The *Encyclopedia Britannica* notes:

Among the most obvious properties of a dilute gas, other than its low density compared with liquids and solids, are the great elasticity or compressibility

A7 There are many useful references that help to introduce this topic more deeply than is possible in this brief exposition. See, for example, Zucker and Biblarz (2002); Hodge and Taylor (1999); Reiner-Decher (1994); Jones and Dugan (1996); Mayhew and Hollingsworth (1996); Moran and Shapiro (2000).

and its large volume expansion on heating. These properties are nearly the same for all gases and ... can be described quite accurately by the following universal equation of state:

$$pv = RT.$$

This expression of behaviour pv = RT is known as the "ideal gas law". Although most gases behave very closely according to this law, there are some small variations from it at low temperatures. In this expression p = pressure, v = "molar volume", T is the "absolute temperature" and R is the universal gas constant.

In order to continue this discussion we need to define some of these basic ideas. The general view of gas behaviour is based on the idea that the molecules of gas in a given contained volume of gas are all moving and colliding and bumping into the walls of the container. Figure A23 is a schematic picture of a gas contained in a rectangular vessel.

The idea of very low temperatures has fascinated physics researchers since it was realised that all matter behaves strangely when temperatures are cooled to near absolute zero. The adopted value for zero Centigrade on the absolute temperature scale is 273.15 K (for Kelvin, a temperature scale named after Lord Kelvin of Largs, also known as William Thompson; 1824–1907, Scottish engineer and physicist), corre-

Figure A23 Schematic picture of a gas. The molecules of the gas move about in their container. For "perfect gases" the motion becomes more rapid as temperature increases and all motion will cease and the "perfect gas" gas will take up zero volume at the "absolute zero" of the temperature scale

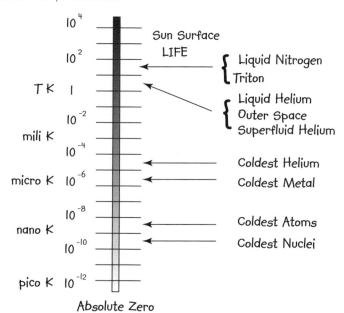

Figure A24 The Kelvin absolute temperature scale (shown logarithmically). Triton is the largest satellite of Neptune and mainly consists of nitrogen ice

sponding to the temperature at which solid, liquid and vapour phases of water are in equilibrium with each other. This "triple point" of water occurs at 0°C or 32°F and 273.15 K (absolute) to make the "ideal gas law" the simple form it is. Although absolute zero (-273.15°C)cannot be reached, temperatures near this value have been attained in laboratory experiments. The scale of temperatures shown in Figure A24 is due to Professor Michael Lea, The Low Temperature Physics Group, University of London.

Molar volume is the volume of "one gram molecule" of a gas, where the gram–molecule measure is the molecular number of the gas in grams. For air, a mixture of oxygen and nitrogen, the average mole is 29 gm and even though it is a mix of two gases, the behaviour of the mixture closely follows the ideal gas law above.

The ideal gas law is a combination of three earlier laws. The first one is Boyle's law, named after the Anglo-Irish scientist Robert Boyle (1627–1691), who in 1662 discovered that, at constant temperature, the volume of a gas is inversely proportional to its pressure. The second law is due to Jaques-Alexandre-Cesar Charles (1746–1823), a French physicist, who discovered that, at constant pressure, a given quantity of gas has a volume directly proportional to its absolute temperature. The third law is that due to Amadeo Avogadro (1776–1856), an Italian physicist, who showed that, at a given temperature and pressure, equal volumes of gases contain the same number of molecules. This number (approximately) 6×10^{23} is called Avogadro's number. For example, the molecular weight of oxygen is 32, and hence 32 gm of oxygen at 0°C (273 K) and atmospheric pressure (usually referred to as "standard temperature and pressure" or STP) contains 6×10^{23} molecules. R is the universal gas constant and has the value 8.3143 J/mole °K.

Now applying the ideal gas law, we can confirm that one gram molecule of oxygen has a "molar volume" of 22.4 litres. Similarly, one mole of air has a mass of 29 gm and has a volume of 22.4 litres. Two other definitions relate to the heat energy needed to raise the temperature of a gas. There are two "specific heats" defined and they are denoted C_p and C_v.

C_p, as the subscript implies is the heat energy required to raise the temperature of a mass of gas by one degree Centigrade (or Kelvin), while the gas is maintained at constant pressure.

C_v is the specific heat defined for the gas at constant volume. These two constants for any gas are related by the universal gas constant as $R = (C_p - C_v)$.

We can now examine the behaviour of compressed air as an energy source. A useful form of compressed air energy source is offered by a plastic 1 litre beverage bottle. With a suitably modified bottle top, the air in the bottle may be pressurised to about two atmospheres, with either a bicycle pump or a small portable electric pump. If now we have a means of releasing the compressed air to do mechanical work, the estimate of available energy is approximated by assuming that the compressed air is exhausted to atmospheric pressure. In that case the work done on the environment by the compressed air is the same as the work done by the environment in compressing the air. Figure A25 is a plot of pressure against volume for a gas. The work done in compressing the air is the area under the graph or

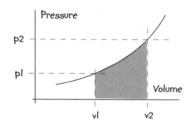

Figure A25 Pressure volume curve for air when being compressed from one atmosphere to two atmospheres

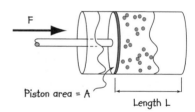

Figure A26 Gas "spring" behaviour

$$w = \int_{p1}^{p2} p\,dv \ .$$

The shaded area under the curve is the work done by the environment on the gas as it is being compressed. We assume the ideal condition that all this work can be recovered as the gas is permitted to expand freely. If the expansion takes place rapidly, as would be the case when the gas is unconstrained, then the expansion (compression) curve is said to be "adiabatic" and the index of expansion (compression) is $\gamma = C_p/C_v$. For air the value of γ is approximately 1.3 and the relationship between pressure and volume is pv^γ = constant.

For the case of 1 litre air compressed to (say) two atmospheres and then permitted to expand into atmospheric air, the values of volume and pressure are:

p_1 = 206,000Pa; p_2 = 103,000;

$v_1 = 10^{-3}m^3$; $v_2 = (p_1 v_1^\gamma /p_2)^{-\gamma} = 1.7 \times 10^{-3}m^3$.

The resulting work done by the expanding gas

$w = p_1 v_1^\gamma [v_2^{1-\gamma} - v_1^{1-\gamma}]/(1 - \gamma)$,

= 101.5 J.

If we take the "very simple" approximation of considering the curve in Figure A25 as a straight line, then we find the shaded area to be

$(v_2 - v_1)(p_1 + p_2)/2 = 108$ J.

Similarly, for a balloon, the work done in expanding from 2.5 kPa above atmospheric to atmospheric pressure is provided by the following data:

p_1 = 105,400Pa; p_2 = 103,000Pa;

$v_1 = 6.4 \times 10^{-3}m^3$; $v_2 = (p_1 v_1^\gamma /p_2)^{-\gamma} = 6.5 \times 10^{-3}m^3$.

The resulting work done by the expanding gas

$w = p_1 v_1^\gamma [v_2^{1-\gamma} - v_1^{1-\gamma}]/(1 - \gamma) = 12$ J.

We can also examine the behaviour of air as a spring by considering a simple experiment using a piston and cylinder arrangement shown schematically in Figure

A26. The air in the cylinder is initially compressed to p_1 and then the piston is moved to compress the air further by decreasing the volume of the contained air. Consider the length of the initial volume to be L and the compression to be (say) ΔL, the area of the piston remaining constant at A.

The increase in force on the piston may be calculated by assuming that the compression is adiabatic. This means that no heat is either absorbed or released by the gas during compression. This is a fair assumption if the compression is small and occurs rapidly.

$$F_1 \quad = p_1 \times A; \ v_1 = A \times L; \ v2 = A \times (L - \Delta L).$$

$$P_2 \quad = (p_1 v_1{}^{\gamma} / v_2{}^{\gamma}) = p_1 [L/(L - \Delta L)]^{\gamma}.$$

$$F_2 \quad = p_2 \times A.$$

$$k \quad = (F_2 - F_1)/\Delta L. \ (k \text{ is the spring constant or elastic modulus of the air spring}) = A \times p_1 [1/(1 - \Delta L/L)^{\gamma}]/\Delta L.$$

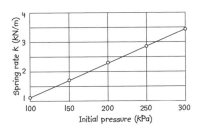

Figure A27 Air as a spring based on the operating arrangement shown in Figure A26

As a result of this expression for the spring constant k of an air spring, the value of k will be a function of the initial pressure in the spring. In this way air, or gas, springs may be modified during operation by increasing or decreasing the pressure within the spring. In some automobiles this effect is used as a means of providing variable rate suspension systems. For L = 100 mm, ΔL = 10 mm and A = 100 mm², the spring constant is plotted against initial pressure p_1 in Figure A27. Clearly, as expected from the equation for k, and indicated in Figure A27, air-spring behaviour is quite linear for small displacements.

A1.4 NEWTON OR ARISTOTLE?

Our understanding of the way mechanics operated in our universe, based on Newton's Laws, has remained virtually unchallenged for nearly 300 years. The Newtonian view of mechanics still remains quite appropriate for most situations, other than those involving speeds close to the speed of light. According to Newton (who, according to Stephen Hawking, was not a very nice man[A8]), bodies moving at constant speed have all the forces acting on them in perfect equilibrium. Moreover, constant acceleration of a body requires a constant (unchanging) force to act on it. Now, these ideas seem relatively easy to understand in simple situations. Surprisingly, these same relatively simple ideas can cause some head-scratching, when introduced into not-so-easy situations. For a considerable time before Newton, the Aristotelian view of the universe of mechanics tended to confuse the concepts of speed and acceleration. Aristotle claimed that the speed of a body at constant acceleration would depend on its mass.[A9] It is interesting to speculate on how students of mechanics would deal with

A8 Hawking (1998).
A9 See the reference to Galileo's experiments on falling masses in Sobel, (2000).

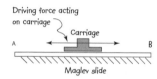

Figure AEI. Maglev test track

less than simple situations, where Newton's Laws should be invoked. Studies of some responses by students, including some university-level participants, by COLOS,[A10] has found that, when faced with these "less than simple" situations, many first-time students of mechanics appear to intuitively adopt the Aristotelian view in favour of the Newtonian view. The following exercises explore the depth of understanding that students have attained in Newtonian mechanics.

AEI. A new type of train is based on the "maglev" principle of floating the carriages on rails using magnetic levitation.[A11] Figure AEI shows a schematic sketch of a rail carriage floating on a length of test track. The following are some questions facing designers of this new form of transport. The conditions of interest to engineers are as follows:

1. F towards B unchanging; 2. F towards B increasing; 3. F towards B decreasing; 4. F towards A unchanging; 5. F towards A increasing; 6. F towards A decreasing; 7. F zero.

Considering that the speeds of the carriage are of order 200 kph (kilometre per hour) and friction and air resistance may be neglected, what will be the speed and acceleration of the carriage (towards A,B, increasing, decreasing, constant) under the seven conditions described?

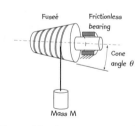

Figure AE2. A new type of fusée

AE2. The fusée, a conical pulley device (similar to that in Figure AE2), was already known to watchmakers in the fifteenth century. It was a device used in mass-driven clocks to equalise the clock spring as it wound down. Although it was a simple device, its main design failing came from the inescapable stretching of the string attaching the mass to the fusée. In modern clocks the fusée was replaced by much more complicated escapement mechanisms. A new invention proposes to use Kevlar or glass fibre strings in a fusée (with virtually zero stretch). It is being tested in a vacuum and friction in the supporting bearing may be neglected. Considering the cone angle θ as the main design variable, the conditions of interest to the designers are:

1. θ positive (as indicated in the figure); 2. θ negative (the cone increases in diameter away from the bearing); 3. θ zero (uniform diameter fusée).

For each case what will be the angular speed of the fusée (increasing, decreasing or constant)?

AE3. A mass M is attached to a string and rotated overhead with the axis of rotation vertical. If the string breaks suddenly, what will be the path of the mass M once released?

A10 Härtel (1994); White (1983).
A11 See www.maglev2004.cn, the official site for the eighteenth *International Conference on Maglev and Linear Drives*, held in Shanghai, PRC, October, 2004.

APPENDIX A2
UNITS OF MEASUREMENT AND CONVERSION FACTORS

In Xanadu did Kublai Khan
a stately pleasure-dome decree,
where Alph, the sacred river, ran
through caverns measureless to man
down to a sunless sea,
so twice five miles of fertile ground
with walls and towers were girdled round…
Samuel Coleridge Taylor 1798

To the inexperienced, though inventive designer, the idea of units of measure may seem unnecessarily pedantic in the context of make-and-test projects. The substantive argument for considering units, and (shock, horror!) conversions between units, is that they form part of the formal engineering language that permits communication between problem specifier and problem solver. In addition, the history of units and their adoption by various institutions and countries provides a fascinating and often illuminating insight into human nature "in the large".

Although difficult to prove, it is likely that even early cavemen used units of measure for exchange (*e.g.*, "my cave is the height of two Triceratops" – about 4 metres, or "I wish to trade five spears of the length of a tyrannosaurus skull" – approximately 1.5 metre). It is certainly true that the compilers of the Old Testament used units of measure as in the case of God instructing Noah *(Genesis 6)*

14   Make thee an ark of gopher wood; rooms shalt thou make in the ark, and shalt pitch it within and without with pitch.

15   And this is the fashion which thou shalt make it of: The length of the ark shall be three hundred cubits, the breadth of it fifty cubits, and the height of it thirty cubits.

A cubit is an ancient unit of length measure used throughout the Old Testament. It originates as early as 3000 BC (about 5000 years ago), and is the distance from the elbow to the end of the middle finger, approximately 454 – 555 mm. By 2500 BC this had been standardised in a *royal master cubit* of about 520 mm. This cubit was divided into 28 *digits* (roughly the width of a finger) which could be further subdivided into fractional parts, the smallest of these being only just over a millimetre.

Early measurements would make use of "natural measures" as in the surveyor's *rod* or *chain*, the grain merchants *bushel* or the precious stone measure of *carat*. The last is derived from the *carob* seeds used in the trading of precious stones, now standardised as 200 mg.

Systems of Measurement and Standardisation[A2.1]

Initially the main types of measures were for *lengths* (distances), *areas* (for example, measuring parcels of land or quantities of cloth), *volumes* (for example, measuring quantities of fluids or grains), *mass* and *time*. As technology changed, other units of measure became necessary, including *temperature*, *heat*, *electric current* or *electric potential* and *illumination* level.

In England units of measurement became standardised in the thirteenth century, though variations (and abuses) continued until long after that. For example, up until 1824 there were three different *gallons* (ale, wine and corn) when the UK gallon, still in current use, was standardised at approximately 277 cubic inches (4.545 litres in SI units).

Initially the United States adopted the English system of weights and measures, with a few minor differences. For instance, the wine-gallon of 231 cubic inches was used instead of the English one of 277 cubic inches. The UK *inch* measured 2.53998 cm whereas the US inch was 2.540005 cm. Both were standardised at 25.4 mm in July 1959.

In France the metric system officially started in June 1799. The unit of length was the metre which was defined as being one ten-millionth part of a quarter of the earth's circumference. The production of this standard required a very careful survey to be done which took several years. To permit proper standardisation of measurements and units the *Bureau International des Poids et Mesures* (BIPM) was set up by the *Convention of the Metre* and has its headquarters near Paris, France.

The *Convention of the Metre*, signed in Paris in 1875 by representatives of seventeen nations and modified slightly in 1921, is a diplomatic treaty which gives authority to the General Conference on Weights and Measures (*Conférence Générale des Poids et Mesures*, CGPM), the International Committee for Weights and Measures and the International Bureau of Weights and Measures to act in matters of world metrology, particularly concerning the demand for measurement standards of ever increasing accuracy, range and diversity, and the need to demonstrate equivalence between national measurement standards.

The Bureau's mandate is to provide the basis for a single coherent system of measurements throughout the world, traceable to the International System of Units (SI). This task takes many forms, from direct dissemination of units (as in the case of mass and time) to coordination through international comparisons of national measurement standards (as in length, electricity, radiometry and ionising radiation).

In all the BIPM recognises seven basic units of measure under the International System of Units, *Système International d'unités*, or SI system, and all other measures and units are derived from these.

A2.1 Source: *www.bipm.org*: a wonderful SI brochure is freely available from this site. See also http://physics.nist.gov/cuu/index.html for the USA version of standard system of units.

The Seven Fundamental SI Units

- *Length* – The *metre* (m) is the length of the path travelled by light in vacuum during a time interval of 1/299,792,458 of a second.

- Mass – The kilogram (kg) is the unit of mass; it is equal to the mass of the international prototype of the kilogram.

- *Time* – The second (s) is the duration of 9,192,631,770 periods of the radiation corresponding to the transition between the two hyperfine levels of the ground state of the cesium 133 atom.

- *Electric current* – The *ampere* (A) is that constant current which, if maintained in two straight parallel conductors of infinite length, of negligible circular cross-section, and placed 1 m apart in vacuum, would produce between these conductors a force equal to 2 x 10^{-7} Newton per metre of length.

- *Temperature* – The *Kelvin* (K) unit of thermodynamic temperature is the fraction 1/273.16 of the thermodynamic temperature of the triple point of water.

- *Quantity of elemental material* – The *mole* is the amount of substance of a system which contains as many elementary entities as there are atoms in 0.012 kilogram of carbon 12. When the mole is used, the elementary entities must be specified and may be *atoms, molecules, ions, electrons*, other particles or specified groups of such particles.

- *Luminous intensity* – The *candela* (cd) is the luminous intensity, in a given direction, of a source that emits monochromatic radiation of frequency 540 x 10^{12} Hertz and that has a radiant intensity in that direction of 1/683 Watt per steradian.

Some Derived Units and Dimensions

All units of measuring physical quantities may be derived from these seven fundamental units. Table A2.1 lists some of the derived units and their symbols. *Area, density, force* and *pressure* are typical derived units. The SI agreed format for all the units and their symbols is in the singular. Typically, ten metre (10 m), or ten kilogram (10 kg), rather than ten metres, or ten kilogrammes. In addition, each quantity has associated *dimensions* in mass (M), length (L) and time (T). These dimensions permit the evaluation of relationships between derived units. Table A2.1 lists the dimension for all the derived units. As an example, the unit of force is Newton, derived by the definition that it is the *force* required to *accelerate* a *mass* of one kilogram at one metre per second per second (invoking Newton's law, $F = Ma$). Its dimensions then become $[MLT^{-2}]$. *Work* is the quantity derived from the definition

Table A2.1 Derived quantities and their dimensions			
Quantity	Unit	Symbol	Dimensions
Area	Square metre	m^2	L^2
Volume	Cubic metre	m^3	L^3
Velocity (speed)	Metre per second	$m\ s^{-1}$	LS^{-1}
Acceleration	Metre per second per second	$m\ s^{-2}$	LS^{-2}
Density	Kilogram per cubic metre	$kg\ m^{-3}$	ML^{-3}
Concentration	Mole per cubic metre	$mol\ m^{-3}$	ML^{-3}
Specific volume	Cubic metre per kilogram	$m^3\ kg^{-1}$	L^3M^{-1}
Current density	Amperes per square metre	$A\ m^{-2}$	$A\ L^{-2}$
Force	Newton	N	MLT^{-2}
Work	joule	J	ML^2T^{-2}
Power (rate of working)	Watt (joule/second)	W	ML^2T^{-3}
Pressure	Newton per square metre, or Pascal	Pa	$ML^{-1}T^{-2}$
Stress	Newton per square metre, or Pascal	Pa	$ML^{-1}T^{-2}$

that *one unit of work is performed* by *one Newton force* moving *one metre*. Hence the units of work are $[ML^2T^{-2}]$ (i.e., force x length). Kinetic energy is the energy of a moving body of mass m at velocity v. Hence $KE = mv^2/2$. The dimensions of KE are $[ML^2T^{-2}]$, or the same as the dimensions for *work*. Clearly, these two derived quantities, *work* and *energy*, have the same dimensions and indeed they are measured in the same units of joule (J).

The dimensions of units also permit the easy interchange between different sets of units. As an example consider the quantity stress expressed in the old civil engineering unit of *kip* (kilo pound per square inch) to be converted to the SI unit of Pascal (Pa = Newton/m^2).

1 pound force = 0.454 kg x 9.8 (the gravitational constant) N,

1 inch = .0254 m,

$$\frac{1000\ lb}{in^2}\left[\frac{[MLT^{-2}]}{[L^2]}\right] = \frac{0.454(kg/lb)\times 9.8\ (\text{gravitational} \bullet \text{acc}^n.)}{.0254^2(\text{metre}^2/\text{inch}^2)}$$

$$= 6894\ kPa\left[\frac{[MLT^{-2}]}{[L^2]}\right].$$

Table A2.2. Prefixes used to denote orders of magnitude in units		
Prefix	Symbol	Power of ten
yotta	Y	24
zetta	Z	21
exa	E	18
peta	P	15
tera	T	12
giga	G	9
mega	M	6
kilo	k	3
hecto[1]	h	2
deca[1]	da	1
		0 $(10^0 = 1)$
deci[1]	d	-1
centi[1]	c	-2
milli	m	-3
micro	μ	-6
nano	n	-9
pico	p	-12
femto	f	-15
atto	a	-18
zepto	z	-21
yocto	y	-24

Note 1. These prefixes are not recommend by SI and are included because they are still in use by some systems of units.

In many situations there is a need to express quantities in very large or very small units. The SI system recommends standard prefixes for denoting various orders of 10 as the multiplier of the basic unit. For example, one kilowatt, denoted kW is 1000 W. Table A2.2 lists the standard prefix set recommended by SI, virtually all prefixes differing by three orders of magnitude (*i.e.*, Giga is three orders of magnitude larger than mega and pico is three orders of magnitude smaller than nano).

There are some important units used internationally that have not yielded to SI standardisation. The following are some examples:

- *knot* = 1 nautical mile per hour = (1852/3600) m/s; The nautical mile is a special unit employed for marine and aerial navigation to express distance. The conventional value given above was adopted by the First International Extraordinary Hydrographic Conference, Monaco, 1929, under the name "International nautical mile". As yet there is no internationally agreed symbol. This unit was originally chosen because one nautical mile on the surface of the Earth subtends approximately one minute of angle at the centre.
- *are* = 100 m², a unit still in common use in the form *hectare*, short for *hecto-are* or 10^4 m².

- *bar* – as a unit of pressure, approximately in multiples of standard atmospheric pressure (*atmospheres* as measured by a *barometer* – hence *bar*). 1 bar = 0.1 MPa = 100 kPa = 1000 hPa = 10^5 Pa.
- *yard* – still used in expressing distances (a mile = 1760 yard); 1 yard = 0.9144 metre; same in the US and UK.
- *pound* – for measuring quantity of substances = 0.453 592 37 kilogram; same in the US and UK.
- *US gallon* (liquid) – volumetric measure = 3.785,411,784 litre.
- *UK gallon* = 4.546,09 litre; different from US gallon.
- *Ton* – a unit of mass = 2000 pound (*short ton*) in US and 2240 pound (*long ton*) in UK.
- Note that the SI unit of megagram (Mg = 10^6 gm) is the *Tonne*.
- The international unit of thermal energy is the calorie (cal) 1 cal = 9.80,665 joule.
- The speed of light in vacuum = 299,752,458 ms^{-1}.
- Acceleration due to gravity = 9.806,65 ms^{-2}.
- Newton's gravitational constant G = 6.674,2E-11 m^{3}kg^{-1}s^{-2}.

Conversion Factors

Quantity	SI unit	To get (other unit)	Multiply SI unit by
Mass	kg	pound (lb)	2.204 62
		ounce (oz)	35.274 0
Length	metre	foot (1/3 yard)	3.280 84
		inch (1/12 foot)	39.370 1
	kilometre (km)	mile (5280 ft)	0.621 37
Volume	litre (l)(=E-03 m^3)	gallon US	0.264 172
		gallon UK	0.219 969
		cubic foot (ft^3)	3.5315E-02
Density	kilogram/metre^3 (kg m^{-3})	pond/cubic foot (lb ft^{-3})	6.242 8E-02
Force	Newton (N)	pound force (lbf)	0.224 809
Moment	Newton metre (Nm)	foot pound force(ft lbf)	0.737 562
Pressure/stress	Pascal (Pa = N m^{-2})	pound force/in^2 (psi)	1.450 4E-04
		pound force/ft^2(psf)	2.088 5E-02
		bar	1 E-05
		millimetre Mercury (mm Hg)	7.500 6E-03
		inch water (in H$_2$O)	4.014 6E-03
	Tonne/metre^2 (Tm^{-2})	UK Ton/ft^2	9.143 789E-05
		US Ton/ft^2	1.024104E-04
	bar	Atmosphere (atm)	0.986 923
Work/energy	joule (J)	British thermal unit (BTU)	9.478 2E-04
		Calorific heat unit (CHU)	5.2657E-04
Power	Watt (Js^{-1})	Horse power (Hp)	1.341 0E-03

APPENDIX A3
SOME PROPERTIES OF PLANE SECTIONS AND ANSWERS TO SELECTED EXERCISES

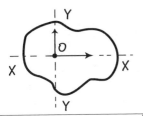

Two relationships of importance are:

Polar 2nd moment of area $I_p = I_{xx} + I_{yy}$, and the Parallel Axis Theorem, $I_z = I_w + Ah^2$, where I_z is the second moment about axis ZZ, I_w is second moment about axis WW, with ZZ and WW parallel and h is the distance between them.

Shape	Area	Centroid O_x	Centroid O_y	I_{xx}	I_{yy}
Rectangle	bh	$b/2$	$h/2$	$\dfrac{bh^3}{3}$	$\dfrac{b^3h}{3}$
Circle	$\dfrac{\pi D^2}{4}$	0	0	$\dfrac{\pi D^4}{64}$	$\dfrac{\pi D^4}{64}$
Triangle	$bh/2 = A$	$\dfrac{(a+b)}{3}$	$h/3$	$Ah^2/18$	$A[b^2 - ab + a^2]/18$
Ellipse	πab	0	0	$\dfrac{\pi ab^3}{4}$	$\dfrac{\pi ba^3}{4}$

ANSWERS TO SELECTED EXERCISES

Doubt is not a pleasant condition, but certainty is absurd.
Voltaire
Everything should be made as simple as possible, but not simpler.
Einstein

E2.1 Newton's bucket

Figure A3.1 Newton's Bucket Experiment

A simple experiment, conducted by Newton, involves somewhat mysterious results. Figure A3.1 shows a bucket which contains water and is suspended by a cord so it is free to rotate around its centreline. The bucket has been turned many revolutions around the centreline so the cord is twisted and exerts a torque on the bucket. The surface of the water is flat before the bucket commences to rotate under the action of this torque and when the bucket is held motionless relative to Earth.

When the bucket is permitted to rotate as the twisted rope unwinds, the water gradually recedes from the middle of the bucket and rises up at the sides of the bucket creating a concave (parabolic) surface as indicated in the figure. Newton's experiment with the rotating vessel of water informs us that even though we do not exert any rotational forces on the body of water (as we would on a mass connected to a string rotating about a centreline), it still experiences centrifugal forces in the Earth's gravitational field. The motion is understood in fluid mechanics as a "free vortex".

E2.2 Einstein's Elevator

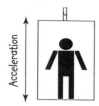

Figure A3.2 Einstein's Elevator

Einstein realised that frames of reference and relative motion are vital concepts. He conducted a thought experiment in which he considered a person in an elevator while the elevator's acceleration was up, down or zero. The effect on the person in the elevator (the observer) was the apparent increase or decrease in mass in spite of there being no change in real mass of the person. If the elevator reaches 9.81 ms^{-2} downward, the observer will appear to be weightless.

E2.3 Schrödinger's Cat

Erwin Schrödinger (1887–1961) was a quantum physicist who used the cat example as an illustration of uncertainty in quantum mechanics. He proposed in 1935 a demonstration of the apparent conflict between what quantum theory tells us is true about the nature and behaviour of matter on the microscopic level and what we observe to be true about the nature and behaviour of matter on the macroscopic level.

First, we have a living cat and place it in a thick lead box. At this stage, there is no question that the cat is alive. We then throw in a vial of cyanide and seal the box. We do not know if the cat is alive or if it has broken the cyanide capsule and died. Because we have no way of knowing if the cat is alive or dead, the uncertainty of its state renders it to be in both states at once until we open the box and discover its actual state.

E2.4 The Bicycle Riders

Analogous to this problem is that of the bicycle riders who start from two towns 40 km apart at exactly the same time riding towards each other. Their average speed is 5 kph. A fly lands on Cyclist I just as he commences the ride, who flicks at the fly which in turn takes off in the same direction as the cyclist. Because the flying speed of the fly is greater than the speed of the cyclist it is able to meet the advancing Cyclist 2. Then the fly returns to Cyclist I, continuing to oscillate between the two cyclists until the two meet. If the flying speed of the fly is 10 kph, how far has it travelled in this process? Clearly, the cyclists meet (on average) in 4 hours from the start (halfway between the two towns) and the fly flies for 4 hours at 10 kph. Hence the distance it covers is 40 km. There is an alternate (long-winded) solution involving an infinite series of distances travelled by the fly.

An amusing story about this analogous problem is one involving John von Neumann (1903–1957), the inventor of the stored program computer. von Neumann was a child prodigy who at age six could divide eight-digit numbers in his head. While at MIT a young colleague posed the bicycle problem to him. von Neumann almost immediately gave the answer. The crestfallen colleague remarked "Ah, you have heard this problem before." "No," said von Neumann, "I simply summed the infinite series."

E2.5 Consider a chess or checkerboard with black and white squares. The opposite two corners of such a board are both the same colour (let us say black). If the covering tile is also in two colours, it will be clearly unable to cover all the squares, because of the 64 board squares there remain 30 black and 32 white squares. Hence if we eliminate all the black–white pairings, eventually there will be a need for a covering tile with two white squares.

E2.6 Approximately 2 minutes per day in error, also represents an error of 8,860 km. The cesium clock can locate objects to within 1 in 3×10^8.

E3.3 Nine dots and six matches.

Figure A 3.3 The Nine Dots and the Six Matches solution

The nine dots and the six matches problems illustrate the nature of delimitation in solving puzzles and problems. Here the solutions really represent cases of "thinking outside the square".

E3.5 Marbles in a bag

Number the bags from 1 to 10 and take one marble from bag 1, two from bag 2 three from bag 3, and so on. If the 9 gram marbles are in bag N then the total mass will be $10 + 20 + 30 + \ldots 9N + \ldots = (500 - N)$ gm.

E6.1 The specific heat of water (Cp) is 1 Calorie gm-1 degree-1. If the temperature of the cup of tea is (say) 80°C, that makes it about 60°C above the average room temperature. If we could recover the whole of this thermal energy from the cup of tea then we could write

$$m.C_p.\Delta T = m.g.h, \text{ or } 1000 \times 60 \times 4.187 = 9.81 \, h$$

$$h = 4.187 \times 1000 \times 60/9.81 = 26 \text{ Km}!$$

An amusing party trick (among a collection of engineers) is to ask some engineer this question and ask someone to "guess" the result.

INDEX